Entrepreneurship Development and Communication Skills

The Authors

Dr. R.R. Chole is Former Head and Director of Extension Education at Marathwada Krishi Vidyapeeth, Parbhani (M.S.) with 36 years total experience in the various capacities as Assistant Professor, Associate Professor, Professor, Chief Extension Education Officer, Registrar, and Associate Dean and Principal in the University system. He has more than 150 research, technical and popular articles and 10 books at his credit including two text books. He is renowned poet of Marathi whose poems have been incorporated in curriculum of M.A. (Marathi) in one of the universities in Maharashtra. He has guided fourteen research students for Ph.D. and thirty seven students for M.Sc.(Agri.) degree. He is recipient of ten prestigious state and national level awards for his significant literary contribution and social work including the award of excellence for his academic contribution.

Dr. P.S. Kapse presently working as Assistant Professor, Department of Home Science Extension Education, College of Home Science, Marathwada Krishi Vidyapeeth, Parbhani (M.S.) with ten years total experience in the extension, research and teaching. He has sixty-five research, technical, popular articles and two text books at his credit. He has designed two practical manuals for undergraduate degree programme. He is recipient of 'The Best Teacher' Award. A paper presented in International Conference at UAS, Dharwad has been judged as 'The Best Paper'. He has qualified NET of ICAR, New Delhi.

Dr. P.R. Deshmukh working for last nine years as Assistant Professor at Department of Extension Education, Marathwada Krishi Vidyapeeth, Parbhani (M.S.) has thirty-five research, technical, popular articles and a text book at his credit and has designed three practical manuals for undergraduate degree programme. He has also contributed to a chapter in a book 'Question Bank – Common Entrance Test for M.Sc. (Agri.)' on Extension Education and guided seven M.Sc. students.

Entrepreneurship Development and Communication Skills

Dr.R.R. Chole

Former Head and Director of Extension Education,
Marthwada Krishi Vidyapeeth,
Parbhani 431 402 (M.S.)

Dr.P.S. Kapse

Assistant Professor,
Department of Home Science Extension Education,
Marthwada Krishi Vidyapeeth,
Parbhani 431 402 (M.S.)

Dr.P.R. Deshmukh

Assistant Professor,
Department of Extension Education,
Marthwada Krishi Vidyapeeth,
Parbhani 431 402 (M.S.)

SCIENTIFIC®
PUBLISHERS

Published by:

Scientific Publishers (India)
5 A, New Pali Road, P.O. Box 91
Jodhpur 342 001 (India)

Branch Office
Scientific Publishers (India)
4806/24, Ansari Road, Daryaganj
New Delhi - 110 002 (India)

E-mail: info@scientificpub.com
Website: www.scientificpub.com

Print: 2026

ISBN: 978-81-7233-793-3

eISBN: 978-93-86347-55-8

Printed in India

PREFACE

Now a day quite frequently people talk about entrepreneurship, both in academic institutions and outside, but in most cases very little attention is paid to important dimensions related to entrepreneurship development. Secondly, many literate people in technologically less advanced areas are grudging accepting entrepreneurship as alternative for employment as it involves risk. However, the economic development of a nation much depends on entrepreneurship development. Therefore different programmes of entrepreneurship development have been initiated at various levels. But in most of these programmes there seems to be lack of comprehensive understanding about the different aspects of entrepreneurship development.

The term Entrepreneurship has got a currency value and is presently at the crest of popularity. With growing education particularly the technical education, a big mass of youth turn to new business ventures. The liberalization of economy, open market system and resultant increase in competition have further added to the value of entrepreneurship as well as role of managers. The education and training of entrepreneurship helps the young entrants understanding and developing their traits and attitudes. It helps them to be competent for preparing business plans, seek financial help from different financial institutions and make them able to manage business on profitable lines. Therefore, the course entitled 'Entrepreneurship Development and Communication Skills' has been introduced in the course curriculum of agricultural universities in India on the recommendation of Fourth Deans Committee. The book contains important aspects of Entrepreneurship Development and Communication Skills leading to building the necessary capabilities of entrepreneurs in the learners.

This book is planned to serve as a text book at undergraduate level for students of agriculture and allied subjects such as horticulture, home science, animal sciences etc. This could also be used as reference book by students at post graduate level in agricultural universities in particular and non-agricultural universities in general. Further it may act as a guide to functional entrepreneurs to inculcate in them the good qualities required for successful entrepreneurship.

Proper utilization of available resources depends upon the abilities, capabilities and potentialities of a person to create successful entrepreneurs through the gate way of multidisciplinary team work. The present work as it happens with all text books, is not the exclusive contribution of the authors but is a derivation of research findings, books and other resources of a large number of academicians, research workers and authors who have been duly acknowledged. Our experience in teaching, training and field extension has helped us to develop get enough material and concepts on the subject and also to provide suitable examples from the field situations.

The aim of adding one more book to continuously expanding list of books on entrepreneurship is to place in the hands of the teacher, students in the agricultural colleges and universities, a handy and simple text book covering complete syllabus. The authors have attempted to clarify most of the fundamental, concepts, principles, techniques and ideas in entrepreneurship development and communication skills with sufficient supporting details.

The book has been divided into seventeen chapters viz. Entrepreneur, Entrepreneurship Development, Motivation and Entrepreneurship Development, Entrepreneurial Behaviour, Entrepreneurship Development Programmes, Setting up of a Small Enterprise, Selection of Product, SWOT Analysis, Market Survey, Formulation of Project, Financial Analysis of Project, Marketing of Product, Government Policy and Institutional Support, Agripreneurship and Indian Agricultural Processing Industry, Communication Skills, Advertising Product and Writing Skills.

We once again duly acknowledge and express our warm indebtedness to the various sources and writers, the views and opinions of whom have been freely incorporated in the book. Thanks to those who motivated and helped us directly-indirectly to bring out this publication. We are grateful to Scientific Publishers (India), Jodhpur not only publishing the present book but also for the earlier text book entitled 'Transfer of Agricultural Technology' which has reached to every nook and corner of India. We hope that the readers will warmly welcome the present work also..

- Authors

CONTENTS

Chapter 1

ENTREPRENEUR

Progress of a nation depends much upon the quality of the human resource that it possesses. Creative, innovative, honest and hardworking people lead the nation in the right direction on the path of overall development of it. However, major part of economic development of a nation depends primarily on the entrepreneurs. Recognizing this fact emphasis is given on development of entrepreneurship. Entrepreneurship is a creative and innovative response to the environment. Entrepreneurs are always in search of new opportunities and they use the opportunities in a skillful manner. Entrepreneurs do new things or do the old things in a new way. The statement of Shiv Khera (1998), the well known thinker, that 'Winners do not do the different things, but they do the things differently' well applies to the entrepreneurs.

Meaning and definition of entrepreneur

The term 'Entrepreneurship' comes from the French word *'entreprendre'* and the German word *'unternehmen'*, both meaning 'to undertake' or 'to do something.' The entrepreneurship process starts when an individual recognizes an opportunity in the environment. The equivalent Hindi word of entrepreneurship is *Udyog* and that of entrepreneur is *Ugyogi* or *Udyogpati*. A number of definitions have been given of an entrepreneur. The economists view him as a fourth factor of production alongwith land, labour and capital. The sociologists feel that certain communities and cultures promote entrepreneurship like for example in India we say that the *Gujratis* and the *Sindhis* are very enterprising, still others feel that entrepreneurs are innovators who come up with new ideas for products, market or techniques. To put it very simply an entrepreneur is someone who perceives opportunity, organizes resources needed for exploiting that opportunity and exploit it. Computers, ATMs, credit cards, mobiles and fast foods are all examples of entrepreneurial ideas that got converted into good and services.

- Entrepreneurs as group of experts who bears the risk of operating a business in the face of uncertainty about the future conditions for taking business advantages. *Frenk Knight (1921)*

- An entrepreneur is one who organizes, operates and assumes the risks in a business venture in an expectation of making a profit. *- Oxford dictionary (1933)*
- An entrepreneur is an economic man who tries to maximize his profits by innovations. He further says that innovations involve problem-solving and entrepreneur gets satisfaction from using his capabilities in attacking problems. *- Haggens E.E. (1970)*
- Entrepreneur as a person who purchases the means of production for combining them into marketable products at profit motive in future or entrepreneur is one who buys factor services at 'certain' price with a view to selling their product at 'uncertain' prices in future. *- Richard Cantillon (1775)*
- Entrepreneur is one who is willing and able to initiate and successfully manage for a length of time an activity that involves at least some degree of personal and organizational risk. *- Khandwalla (1979)*
- Entrepreneur is one who initiates and establishes an economic activity or enterprise. *- Pareek and Nadkarni (1978)*
- The person using modern devices and equipment could be considered as an entrepreneur, provided he is innovative, he is influenced by the urge to take risk and has been acting on the spur of intuition which proves to be right afterwards. *– Gaikwad (1978)*
- Joseph Schumpeter (1977) was first to introduce dynamic concept of the entrepreneurial function and was major writer to put the human agent at the centre. According to him, 'the entrepreneur is a dynamic agent of change, is the catalyst who transformed increasing physical, natural and human resources in a corresponding production possibility.' An entrepreneur as an innovator with potentialities of doing things in a new way. Entrepreneurship, on the other hand, is a dynamic process with *'innovation'* and *'risk bearing'* as its two basic elements involving multiplicity of activities towards establishment of an enterprise.
- Entrepreneurs are people who have the ability to see and evaluate business opportunities, to gather the necessary resources, to take advantage of them and to initiate appropriate action to ensure success. *-International Labour Organization (1982)*
- An entrepreneur creates something new with value by devoting the necessary time and effort, assuming the accompanying financial, psychic and social risks, and receiving the resulting rewards of monetary and personal satisfaction and independence. *– Robert Hisrich and Michael Peters*

- Entrepreneur is one who exercises some control over the means of production and produces more than what he can consume in order to sell it for profit. He executes the functions of coordination, organization and supervision. Entrepreneur is basically an innovator who introduces new combinations. An entrepreneur is a person with a high need for achievement. He is energetic and a moderate risk taker. *- Mc Clelland*
- An entrepreneur searches for change, responds to it and exploits opportunities. Innovation is a specific tool of an entrepreneur; hence an effective entrepreneur converts a source into a resource. *- Peter Drucker (1970)*
- An entrepreneur as one, who takes initiative, accepts risks of failure and has an internal locus of control. *- Albert Sharpero*
- Entrepreneurship is the character, practice and or skill that combines innovativeness, readiness to take risk, sensing opportunities, heightened initiative, perceiving and mobilizing potential resources and concern for standard of excellence, persistence in achieving the goal, positive orientation to problem solving and constant strive for growth and development. When all these attributes are developed in one person, the person becomes an entrepreneur.

Conclusions

1. On the basis of definitions given by different authors, it may be concluded that the entrepreneur is perceived as an individual with certain characteristics helpful in conceiving, initiating, establishing, running and finally managing an enterprise. An enterprise may vary from starting a small shop to establishing an advanced technology based industry.
2. Entrepreneur is a person who gets things done, starts a venture on his or her own and is able to create something that produces an outcome such as wealth.
3. Entrepreneur introduces something new into the economy, which may be a new product or process, or finds a new market for a product or process already known.
4. Entrepreneur is an innovative person undertaking an enterprise with calculated risk for maximizing profit.

Meaning and definition of entrepreneurship

The concept of entrepreneurship has been visualized by various authors in different ways. Some of the definitions are given here below.

- Entrepreneurship is a process of change comprising three behavioural components such as (i) identification, evaluation and exploitation of an opportunity, (ii) management of a new or transformed organization so as to facilitate production and marketing of new goods and services and (iii) creation of value through successful exploitation of new idea, i.e., innovation.

  - Bahal (2008)

- Entrepreneurship comprises any purposeful activity that initiates, maintains or develops a profit-oriented business in interaction with the internal situation of the business, or with the economic, political and social circumstances surrounding the business. *- Cole, A.H. (1949)*

- Entrepreneurship is the function of seeking investment and production opportunity, organizing an enterprise to undertake a new production process raising capital, hiring labour, arranging the supply of raw materials, finding site, introducing a new technique and commodities, discovering new sources of raw materials and selecting top managers of day to day operations of the enterprise. *- Haggen (1970)*

- Entrepreneurship can be described as a creative and innovative response to the environment. Such response can take place in any field of social endeavour-business, industry, agriculture, education, social work and the like. Doing new things or doing things that are already known in a new way is, therefore, a simple definition of entrepreneurship. *- Rao and Mehta (1978)*

- Entrepreneurship is the process of creating something new with value by devoting the necessary time and effort, assuming the accompanying financial, psychic, and social risks, and receiving the resulting rewards of monetary and personal satisfaction and independence. Undertaking of an enterprise is entrepreneurship.

  - Robert Hisrich and Michael Peters

- Entrepreneurship is the dynamic process of creating incremental wealth. This wealth created by individuals who assume the major risks in terms of equity, time and / or career commitment of providing value for some product or service. The product or service itself may or may not be new or unique but value must somehow be infused by the entrepreneur by securing and allocating the necessary skills and resources. *- Robert Ronstad*

- Entrepreneurship begins with action, creation of new organization. This organization may or may not become self sustaining and in fact, may never earn significant revenues. But, when individuals create a new organization, they have entered the entrepreneurship paradigm.

It may be summarized here that although each of these definitions of entrepreneurship has slightly different perspective, they all contain similar notions such as newness, organizing, creating wealth and risk taking. Entrepreneurship is found in all professions, including education, medicine, law, agriculture, research, social work and engineering. From this view point, the definition given by Hisrich and Peters is more appropriate as it includes all types of entrepreneurial behavior. Entrepreneurship is the set of activities performed by an entrepreneur. Therefore, entrepreneur precedes entrepreneurship.

In short, entrepreneurship is an operational process initiated by an entrepreneur, by which he/she reaches to ultimate objective of establishing an enterprise and manage it profitably. This entire process can be termed as 'Entrepreneurship' which has been shown in the following way.

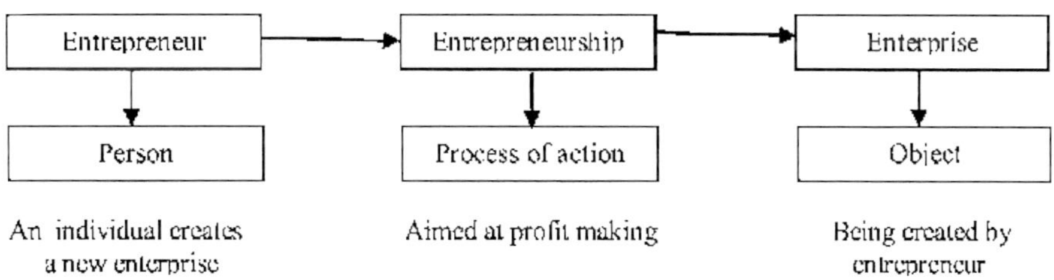

Figure 1: Relationship among entrepreneur, entrepreneurship and enterprise.

Characteristics of an entrepreneur

The different definitions of entrepreneur mean to say that entrepreneurs are individuals with distinct characteristics helpful in conceiving, initiating, establishing, running and managing an enterprise successfully. An entrepreneur therefore may be differentiated not only in terms of activities he or she undertakes but in terms of his or her lifestyle, attitude, values and overall behaviour which together make the personality of entrepreneur different than an ordinary person.

In other words, the entrepreneurs are known for their special characteristics which make them entrepreneur. An entrepreneur may not possess all the characteristics mentioned here, but may possess own set of characteristics. It may be noted that there is no sex discrimination in becoming an entrepreneur; males and females are equally competent to become entrepreneur in any field.

1. **Risk taking:** Risk means the condition of not knowing the outcome of an activity or decision. A successful entrepreneur prefers a situation where there is moderate challenge, which can be overcome by his efforts. He prefers to take calculated risk in

dealing with a challenging situation. He enjoys the excitement of challenge, but he does not gamble. This characteristic is one of the most important. Success or failure of business depends on how decisions are made. A risk situation occurs when one is required to make a choice between two or more alternatives whose potential outcomes are not known. A risk situation involves potential gain or loss. The greater the possible loss, the greater is the risk involved.

2. **Creativity and innovativeness:** Entrepreneurs always try different alternatives to achieve their goals at their disposal. They tend to be creative and innovative. For them, the method of choice is the method, which will work best. They are ingenious at adopting and modifying whatever is at hand to solve the problems or achieve the objective.

3. **Achievement motivation:** Entrepreneurs have high achievement motivation, i.e., an urge for excellence and desire for success in competing with standards set by others as well as one sets for oneself.

4. **Personal drive and energy:** Entrepreneurs can work for long hours for days together with minimum basic amenities.

5. **Self confidence:** The entrepreneurs believe that they are capable of doing things and have the ability to control the actions and activities of other people. They have confidence in their own capabilities to achieve the goals set by them. They also believe that events in their lives are mainly self-determined, and they have major influence on their own destinies and have little faith in fate.

6. **Taking initiative:** The entrepreneurs are independent and highly self-reliant persons. Entrepreneurs actively seek and take initiatives. When they undertake a task, they make it sure to complete it. They prefer situations where the result depends on their ability or efforts rather than a chance or other factors beyond their control.

7. **Seeking feedback of own efforts:** Entrepreneurs are very concerned about their performance. They prefer to gauge their achievement objectively and take corrective action immediately. Constant feedback is stimulating and satisfying to them and they use it to modify and improve their efforts at appropriate levels.

8. **Internal locus of control and seeking personal responsibility:** Successful entrepreneurs like to deal with situations where they are personally responsible for the success or failure of the operation. They are willing to take credit for the

success, but are equally prepared to accept blame in situation of failure. The entrepreneurs do not believe that the success or failure of new venture depends mostly upon luck or fate, or other external, personally uncontrollable factors. Rather, the entrepreneurs tend to believe that one's personal accomplishments as well as setback lie within one's personal control and influence.

9. **Enduring involvement:** This characteristic differentiates between one entrepreneur and a promoter. They make commitment for a long term project and to working towards goals that may be quite distant in the future. Successful entrepreneurs also know building of a business requires a total treatment to attain a distant goal.

10. **Orientation towards future goals:** Entrepreneurs are very goal oriented. They have an ability to set goals and commitment to work towards them. Entrepreneurs are goal and action oriented. They have high need for achievement. They set clear, measureable goals and accordingly they set priorities, measure and guide their time allocations.

11. **Money as measure of performance:** Money is the tool to judge the performance. Profits, gains and net worth are seen as measures of how well an entrepreneur is doing. The entrepreneur is always involved in process of making money, going out and investing it in another venture and then starting all over again.

12. **Systematic planning:** Planning is really nothing more than decision-making, i.e., deciding what to do, how to do and when to do. The entrepreneur is the most knowledgeable person to chalk out the business plan. Although he can take help from outsiders but ultimate responsibility lies on the entrepreneur himself. An entrepreneur believes in planning and takes logical steps to achieve the goals.

13. **Dealing with failure:** Failure does not deter an entrepreneur, rather he/she analyses the failure as feedback information, and take appropriate corrective steps. Entrepreneurs are oriented towards problem solving. Good entrepreneur converts failures into opportunities.

14. **Information seeking:** An entrepreneur possesses information seeking behaviour and is always hunting for relevant current information related to the product, process, market, *etc.,* to make sound decisions.

15. **Leadership:** An entrepreneur follows a style of persuasive leadership which motivates the team members to put in their maximum effort. He has an ability to organize and integrate efforts of team and lead them to predetermined orals.

16. **Utilization of resources:** Successful entrepreneurs are always aware of the resources, outside as well as inside and understand when and how to utilize them efficiently. They seek expertise and assistance in achieving their goals. They are not so egoistic in purely individual achievement of goals and independent accomplishment that they will not seek aid from other or refuse others offer of help.
17. **Social networking:** Developing a good network of relationships with the persons, agencies, organizations connected with the enterprise is an important characteristic of a successful entrepreneur.
18. **Persistent problem solving:** Successful entrepreneurs are not afraid by difficult situations. They possess intense level of determination and desire to overcome blocks. While entrepreneurs are extremely persistent, they are also realistic in recognizing their own abilities, and from where they can get help to perform difficult but necessary tasks.
19. **Set high standard for themselves:** High performing entrepreneurs also possess internalized competitive spirit in which they continuously engage in competition with themselves to beat their last best performance.
20. **Adventurous:** Successful entrepreneurs are not satisfied with what is; nor are they content with what they have achieved. Their standards are high, and after they attain their goal, they set new high goals and strive for its attainment.
21. **Tolerance to ambiguity:** Successful entrepreneurs are better able to tolerate ambiguity and uncertainty. They do not require a complete structuring of a situation in order to function and in making of decisions. Entrepreneurs less prefer job security and permanency than other managerial counterparts.
22. **Other characteristics**
    a. Visionary
    b. Able to overcome adversities
    c. Optimistic
    d. Skill in human relations
    e. Effective communicator
    f. Decision maker
    g. Motivator
    h. Organizer

(Singh *et al.*, 2006)

Role demands of entrepreneur

A set of roles need to be performed by an entrepreneur to attain goals. Following are the role demands of an entrepreneur.

1. Accommodation to the venture

- Must be prepared to 'give all'.
- Particularly in early stand up years.
- Crucial decisions – Marriage, raising family, community involvement.

2. Total immersion and commitment

- Launching business requires total immersion and commitment.
- Not a part time job.
- Managerial role could be delegated, but entrepreneurial role cannot be.
- According to Ned Hiezer : You can hire managers, they are a dime dozen, but you can't hire an entrepreneur.

3. Creativity and innovation

- Entrepreneur's career anchor is creativity and innovation.
- Managers find their career anchor in competency and efficiency.
- College professors prefer autonomy as a career anchor.

4. Knowledge of the business

- Must have thorough and proven operating knowledge of the business.
- Formal education has not been a distinguishing requirement.

5. People and team building

- High potential ventures are rarely sole proprietorship.
- Entrepreneurs must build a high quality team.
- For raising the capital and growing the business team building is essential.

6. Economic values

- Must share key values of free enterprise system – private ownership, profits, capital gains, responsible growth.

- However should not exclude social or other values.
- Personal values have profound effect on development of team and business value – The value of anything is what it is worth to someone, that is, what it will produce for him or briefs about what is desirable or undesirable.

7. Ethics

- Ethics – Principles of conduct which help us judge whether a choice or action is good or right.
- Business community has its own ethical standards.
- Entrepreneur tends to possess situational ethics.
- Personal ethics of entrepreneur are decided by the needs and demands of the situation.

8. Integrity and reliability

- Integrity and reliability are must for aspiring entrepreneurs.
- Reputation for dependability, reliability and honesty is essential.

Identifying Potential Entrepreneurs

Success of small ventures depends on

1. Inherent viability of the project.
2. Project planning.
3. Project implementation.
4. Project management.

Entrepreneurial competency

1. Knowledge - Refers to collection of facts, values, information, etc.
2. Set of skills - Ability to demonstrate a sequence of behaviour in attaining a performance goal.
3. Appropriate personality – It is comprised of desired motives and traits.
4. Motives – An urge for which one has continuous concern that directs one towards certain action to satisfy the concern.
5. Traits – Characteristics way in which person responds to set of stimuli. It includes thoughts and psychomotor activities.
6. Entrepreneurial competency that an individual possesses to perform a given task effectively is a combination of all these characteristics.

7. Existence of these characteristics may or may not be known or may be unconscious aspect of the person.

Identifying Entrepreneurial Competencies

Soft and hard skill competencies

1. Soft skill competencies are those which do not constitute professional knowhow and are functionally interchangeable.

   For example – Decision making skills, social relationship skills, problem solving skills, persuasive skills, skillful use of influence strategy, orientation to do things faster.

2. Hard skills are those related to given profession and are acquired through education or hand on experience in performing task.

   For example – Designer engineer engaged in designing gears, Agriculture engineer designing a bullock cart.

Job Competencies of Entrepreneurs

Entrepreneurship Development Institute of India conducted a study under the guidance of David McClelland, a reputed behavioural scientist. It was found out that possession of certain competencies or abilities result in superior performance. An entrepreneur may possess certain competencies and at the same time it is possible to develop these through training, experience and guidance. Various competencies required for superior performance were identified during the study and are as under:

1. **Initiative:** It is the entrepreneur who takes or initiates the first move towards setting up of an enterprise. It basically is an innovator who carries out new combinations to initiate and accelerates the process of economic development. Takes action that go beyond job requirements or the demands of the situation. He does things before being asked for or forced by events. Entrepreneur acts to extend the business into areas, products or services.

2. **Seeing and acting on opportunities:** An entrepreneur is always on the look out or searching for opportunity and is ready to exploit it in the best interests of the organization. Looks for & takes action on opportunities. Sees and acts on opportunities - business, education, personal growth. Seizes unusual opportunities to obtain financing equipment, land, work space or assistance.

3. **Persistence:** An entrepreneur is never disheartened by failure. He believes in the Japanese proverb 'Fall seven times, stand up

eight'. Takes repeated action to overcome obstacle that gets in the way of reaching goals. Takes repeated action or different actions to overcome an obstacle that gets in the way of reaching goals. Takes action in the face of a significant obstacle.

4. **Information seeking:** A successful entrepreneur always keep his eyes and ear open and receives new ideas which can help him in realizing his goals. Takes action on own to get information to help reach objectives or clarify problems. Does personal research on how to provide a product or service. Consult experts for technical advice, personally undertakes research, analysis, or investigation.

5. **Commitment to quality:** Successful entrepreneurs are concerned with high quality work and do not believe in moderate or average performance. They set high quality standards for themselves and then put in their best for achieving standards. Entrepreneurs act to do things that meet or beat existing standards of excellence. They state a desire to produce work of high quality and compare own work or own company's work favourably to that of others.

6. **Commitment to work contract:** Successful entrepreneurs are prepared to make all sacrifices for honouring the commitments they have made. Places the highest priority on getting a job completed. Makes a personal sacrifice or expends extraordinary effort to compete a job. Accepts full responsibility for problems in completing a job for others. Pitches in with workers or works in their place to get the job done. Expresses a concern for satisfying the customer.

7. **Efficiency orientation:** Top performers are always keen to devise new methods aimed at promoting efficiency. They are keen to evolve and try new method aimed at making working easier, simpler, better and economical. Finds ways to do things faster or with fewer resources or at a lower cost. Entrepreneurs look for or find ways to do things faster or at less cost. Uses information or business tools to improve efficiency.

8. **Systematic planning:** Successful entrepreneurs develop or evolve future course of action keeping in mind the goals to be realized. An entrepreneur develops and uses logical, step-by-step plans to reach goals. Plans by breaking a large task down into sub-tasks. Develops plans anticipating obstacles. Find and evaluates alternatives.

9. **Problem solving:** Successful entrepreneur takes problem as a challenge and put in their best for finding out the most appropriate solution for the same. Identifies new and potentially

unique ideas to reach goals. Switches to an alternative strategy to reach a goal. Generates new ideas or innovative solutions to a problem.

10. **Self confidence:** Top performers are not cowed down by difficulties as they believe in their own abilities and strengths. They have full faith on their knowledge, skill and competence and are not worried about future uncertainties. Has a strong belief in self and own abilities. Expresses confidence in own ability to complete a task or meet a challenge. Sticks with own judgment in the face of opposition or early lack of success. Does something that he says is risky.

11. **Assertiveness:** An assertive person knows what to say, when to say, how to say and whom to say. He believes in his abilities and ensures that others fall in line with his thinking, aimed at promoting the interests of the organization. Confronts problems and issues with others directly. Tells others what they have to do. Reprimands or disciplines those failing to perform as expected.

12. **Persuasion:** A successful entrepreneur through his sound arguments and logical reasoning is in a position to convince others to do the work the way he wants them to do. Successfully persuades others and convinces someone to buy a product or service. Convinces someone to provide finance. Convinces someone to do something else that he would like that person to do. Asserts own competence, reliability or other personal or company qualities. Asserts strong confidence in own company's or organization's products or services.

13. **Effective monitoring:** This is to ensure to the performers that everything is carried out in their organizations as per their wishes. Entrepreneur develops or uses procedures to ensure that work is completed in time and that work meets standards or quality. Personally supervises all aspects of a project.

14. **Concern for others welfare:** A successful entrepreneur tries to promote organization's interests through promotion of interests of the workers. He takes personal interest in solving problems confronting workers and generates the feeling that there is independence of the interests of workers and the management. Takes action to improve the welfare of employees. Takes positive action in response to employee's personal concerns. Expresses concern about the welfare of employees.

15. **Use of influence strategies:** A successful entrepreneur possesses the ability to evolve and effective relevant strategy, aimed at safeguarding or promoting organization's interest. Use

a variety of strategies, to affect others. Acts to develop business contacts. Uses influential people as agents to accomplish objectives. Selectively limits the information given to others. Uses a strategy to influence or persuade others.

James J. Berna has stated following competencies found in a good entrepreneur.

a. He is an enterprising individual.

b. He is interested in advancing technology and in improving the quality of his product.

c. He is interested in expanding the scale of his operations and reinvest earnings to this end.

d. He is energetic, resourceful, alert to new opportunities, able to adjust to changing conditions and willing to assume risk involved in change

Chapter - 2

ENTREPRENEURSHIP DEVELOPMENT

Entrepreneurship development is a process involving a number of phases and steps. The process has been viewed at two levels – one at the level of individual - developing entrepreneurial qualities in an individual and another at the level of community - creating an environment in which entrepreneurship activities can grow. Experience has shown that two major factors have played significant role in developing entrepreneurship. One of them is the development of human factor, i.e., entrepreneur himself. Another major factor is the development of environment where entrepreneurial activities can flourish and grow. The process of entrepreneurship development at both community and individual level has been discussed in this chapter.

[A] Process of Entrepreneurship Development at Community Level

Many attempts at accelerating the growth of entrepreneurship in different areas have yielded a variety of experiences. Some were highly successful; others disappointing and perhaps the majority were the cases that had the possibility of considerable success, if certain improvements were made. Analyzing these past experiences mainly from the perspective of the client system, the whole process of entrepreneurship development, (especially in a target group that has yet to imbibe the entrepreneurial value in any significant measure) can be seen as a process consisting of three broad phases.

1. Stimulatory Phase

The stimulatory activities help in the emergence of entrepreneurs in society. They prepare the background where from entrepreneurship can sprout. People start looking for entrepreneurial pursuits. These generate the initial entrepreneurial motivation; help people perceive incentives and also offers opportunity to acquire relevant information and skill. All these taken together help stimulate entrepreneurship in a society. This initial phase would include all such activities that stimulate efforts such as:

1. Generating entrepreneurial awareness in the community through well-planned publicity.
2. Identifying and selecting potential entrepreneurs.
3. Helping them through training to raise their motivational level.
4. Improving their knowledge and skills in modern management methods.
5. Developing technical competence relevant to the product or service selected.
6. Helping them to develop project report.
7. Making available techno-economic information and project profiles.
8. Entrepreneurs to select new product.
9. Developing a data bank on new products and process available to the target group.
10. Evolving new products and processes relevant to local situation.
11. Public recognition of entrepreneurial excellence.

2. Support Phase

The support activities provide infrastructure facilities, resources, abilities and skills to entrepreneurs for enterprise launching and management. These activities provide nurturance and helps the already stimulated entrepreneur to move ahead to reach his goal. This includes all such activities that help entrepreneurs in establishing and running their enterprises. These activities remove many hurdles, which are likely to cause sickness or discourage the new entrepreneurs. The activities in this phase may include:

1. Registration of the unit.
2. Arranging finance of any type such as fixed capital and or working capital.
3. Providing land, shed, power, water, etc. for establishing the unit.
4. Guidance for selecting and obtaining plant and machinery and layout.
5. Helping in purchase of plant and machinery.
6. Issue of license for scarce raw materials.
7. Providing common facilities.
8. Granting tax relief and other types of subsidy.
9. Offering management consultancy.

10. Assisting in marketing the product either through reservation or assigning government quota purchase, etc.
11. Providing information related to industry.

3. Sustaining Phase

Sustaining phase activities refer to all such efforts that facilitate growth and continuity through expansion, modernization, diversification and technology upgradation of on-going enterprises and rehabilitation of sick units. This is a phase in which entrepreneurs are supported to continue their enterprise in efficient and profitable manner.

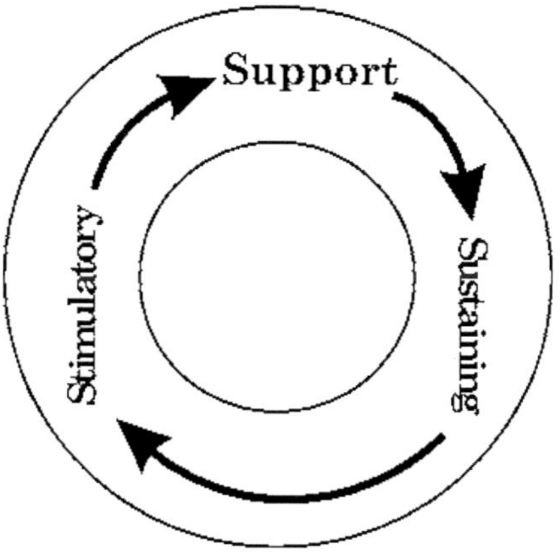

Figure 2 : Cycle of entrepreneurship development process.

After the stimulation of entrepreneurship and subsequent adequate support, the entrepreneur succeeds in starting his unit. Quite often, however an entirely new set of problems crop up before him once he commences production. He may not be fully prepared for some of them. It may be non-profitable and non-viable due to many unforeseen and uncontrollable factors. The small industries are susceptible to a variety of vagaries. Even well managed units may suffer closure for want of raw materials, lack of availability of substitutes, change of skilled hands, *etc.* Such casualties are comparatively higher in the small units and require a lot of sustaining activities by the promotional agencies. The sustaining activities may include:

1. Helping in modernization, diversification, and expansion or product substitution.
2. Additional financing for full capacity utilization.
3. Deferring repayment or interest depending on the situation.
4. Help and guidance in diagnosing the cause of failure or low production or profit.
5. Modification or change in legislation or policy affecting marketing.

6. Quality testing and quality improvement at low cost.
7. Creating common facilities, which are not feasible in a single unit but are needed by the unit both in production and marketing.

The process of entrepreneurship development at community level involves three phases as described above. However the process does not end with sustaining phase. It is rather a continuous cycle. With new needs as well as modification or substitution of a product, further stimulation and support is needed.

Balanced attention to three phases

Each phase of the activities is highly interacting, supplementary and crucial to the other phases. In absence or negligence of any phase, the other phase may be rendered in failure. Often it is found that the promotional agencies place excessive emphasis on support activities such as arranging finance, providing plant and machinery or long term loan and establishing industrial estates to provide the infrastructure like land, shed, power, etc. However, in entrepreneurship development process, support phase though important, lies between stimulatory and sustaining phases. If a programme of development of entrepreneurship ignores the stimulatory phase and begins straightway with support activities, the outcome will be of two types – non-availability of entrepreneurs or availment of facilities by outsiders. In the situation when there is non-availability of entrepreneurs, and if further efforts are made so that outsiders can come and avail the existing facilities then the result will be very high rate of sickness which will be further aggravated if the programme does not include sustaining activities.

Stimulatory activities have sometimes been over emphasized without ensuring a matching degree of support and sustaining activities. In such cases, it has been observed that the supply of entrepreneurs is accelerated and bulks of them are produced in the society. But soon they are frustrated due to lack of timely supportive response from concerned agencies. This not only demotivates them, but also has a negative multiplier effect - it makes even others who might be wanting to come forward demotivated and skeptical. The ultimate casualty in this process is the credibility of the agencies concerned with entrepreneurship development. Similarly, if provision is made for sustaining activities without arranging a matching degree of stimulatory or support activities, there may not be sufficient number of units to be sustained.

[B] Process of Entrepreneurship Development at Individual Level

The human factors refer to the attitude, desire and motivation of an individual, his capability to perceive the environmental changes and opportunities as well as his ability to solve the problems, which he is likely to face. The training is effective in developing all these aspects of human factors provided it is planned well with balanced emphasis on all the aspects. Training has played crucial role for all such strategies in initiating and accelerating the process of entrepreneurship. Entrepreneurship development requires career building that results in emergence of people opting for entrepreneurship.

An analysis of the entrepreneurship development process reveals that it follows a sequence of development of individual-personality, capabilities and abilities.

The first generation of entrepreneurs requires developing following things:

1. Entrepreneurial quality and motivation;
2. Capabilities for enterprise launching and resourcing;
3. Ability for enterprise management; and
4. Sense of responsibility to the society that promotes and supports them.

1. Entrepreneurial quality and motivation

Generally, we find people opting for wage-earning career. This tendency spreads and popularizes such orientation. Social institutions such as the family as well as school, more or less support development of qualities like conformity and compliance, which are not conducive to the growth of entrepreneurial values. As a result of this, creativity, risk taking, perseverance, innovativeness and problem solving orientation, which are some of the accepted entrepreneurial qualities, are not encouraged. Whenever efforts are made to induct people to entrepreneurial career, such entrepreneurial qualities are generally found lacking or dormant. These are required to be aroused to an extent that people may start opting for entrepreneurial career. This is a basic requirement and much needed force, which drives people to their new ventures. Left to themselves, such qualities and motivation will be developed only in very few. Accordingly, as a part of the planned programme of entrepreneurship development, the inputs have to be thought out and administered effectively to ensure development of minimum entrepreneurial qualities as well as motivation adequate enough to drive them to entrepreneurial pursuits.

2. Capability of enterprise launching / resourcing

This opportunity exists in the society but all of us are not sensitive to it. Most of us in the society can perceive only the apparent and traditional openings for earnings. Similarly, resources are also available but very few make the efforts to make use of them. Together with economic insight for sensing opportunities in the area, prospective entrepreneurs may have to develop capability for selecting suitable project, formulating project report, arranging plant machinery, *etc.*, and availing facilities and resources relevant to the launching of their enterprises. These are to be developed through training interventions.

3. *Ability for enterprise management*

The enterprise may be small or big, demands good management abilities in its owner/manager. Several of management such as production/ components, marketing, financial management, etc. are crucial for an enterprises. These have direct influence on the results and, are therefore, necessary determinants for sustenance of an enterprise. The management inputs to the potential entrepreneurs also raise their expectancy for success. However, the intensity of these inputs may vary depending upon the size of enterprise that is selected by an entrepreneur / self-employed person.

4. *Sense of responsibility to society*

The entrepreneurs who are developed and promoted at the social cost have certain responsibility to the society that promotes and supports them. The government and other public institutions that invest on them also expect something in return. In order to sustain the efforts of developing entrepreneurship, it is necessary that a sense of responsibility towards the society in general and towards the entrepreneurial movement in particular needs to be developed among the potential entrepreneurs, especially those belonging to the younger generation. These entrepreneurs need to follow certain discipline, which is useful for entrepreneurial career. Such discipline may cover subjects like repayment behaviour, response to tax and statutory requirements, progressive outlook towards labour and above all care for ecology and environment. These aspects related to individual behaviour, abilities and capabilities follow a logical sequence of development, which ultimately drives prospective youth to actual self – employment / entrepreneurial career.

Stages in Entrepreneurial Growth Process

In the context of employment generation, the three terms, i.e., entrepreneurship, self-employment and income generation are often used interchangeably. Although there are lot of commonalities among these concepts, yet the three terms are not the same (or at overlapping concept).

Self-employment refers to full time involvement in his or her own occupation or pursuits in which one may or may not have to take the risk, to mobilize inputs and other resources, to organize total productions and services or to care for marketing the product and services. Income generating activities on the other hand are often part time, casual and are practiced for the purpose of raising additional income. The functions of entrepreneurs as explained earlier thus suggest that all entrepreneurs are self-employed and income generating persons. But all self-employed and income-generating persons may not necessarily be entrepreneurs. Similarly, all self-employed persons are income-generating persons, but all income generating persons may not necessarily be self-employed. Accordingly, all the three may be useful means for employment generation, but every term varies in their scope and impact. The three, however, can be viewed as 'initial', 'middle' and 'terminal' stages in 'entrepreneurship growth process' at individual level.

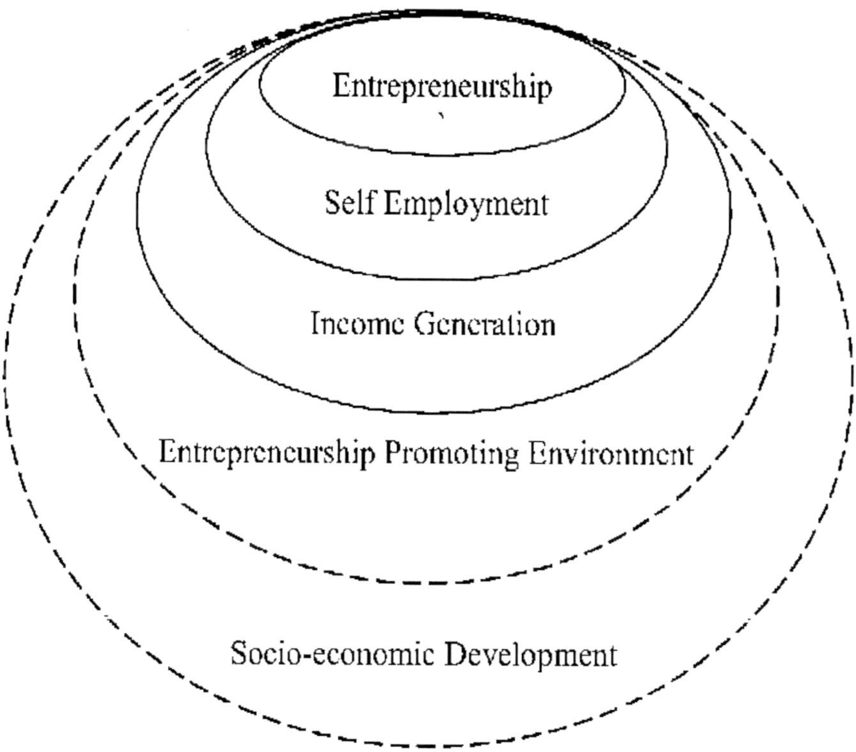

Figure 3: Entrepreneurship growth process.

Each stage may be relevant to different category of society depending on their level of economic development and previous experiences. Each stage contributes significantly in the movement towards entrepreneurship. Income generation experiences encourage self-employment and self-employment efforts facilitates graduating to entrepreneurship.

Differences among income generating activities, self employment and entrepreneurship

Income generating activities	Self employment	Entrepreneurship
• Supplementary income • Partial employment • Usually collective • Managed by NGOs / SHGs • Most common development activity • Programs stop usually at skill training • Programs generally under schemes • Target (number) oriented • Group employment • Seldom result in real income generation • Last as long as the grant • Low investment • Useful employment oriented • No risk	• Income from one's own input • Full time employment • Usually not collective • Sometimes part of SHGs • Fairly common development activity • Mainly artisan skills used own occupation • Programs generally schematic • Target (number) oriented • Employment for self only • Not growth oriented • Lasts longer but limited • Low investment • Traditional occupation oriented • Almost no or less risk	• Profit • Full time employment • Individual, partners, group • Managed by the owners, partners, etc. • Uncommon development activity • Needs more complex skills-Business skills • Few programmes • Programmes focused on economic viability • Employs more than one person • Economic growth oriented through value addition • Once started, it is a continuous process • Investment may be high or low • Market oriented • High risk

Though the above three stages in the entrepreneurial growth process have been well recognized, the authors would like to add two more stages, *viz.,* entrepreneurship promoting environment and socio-economic development. This would serve two purposes, the entrepreneurship development at individual level and societal or community level will be meaningfully combined and the process of entrepreneurship development will be fully explained. Therefore, two dotted circles have been added in Figure 3. The success obtained by an entrepreneur in terms of self-employment and income generation motivates potential entrepreneurs in the community creating environment conducive to entrepreneurship development. Promotional activities by government and other agencies lead to success and more number of enterprises would be established. Lastly, this process results into socio-economic development of the community as a whole through employment generated in the web of enterprises established.

Need of Entrepreneurship

According to Sridharan (2002) the need of entrepreneurship development is as under:

1. **For accepting challenging career option:** After completion of formal education, broadly one has two career options. One can employ oneself in government service, public and private sectors, accepting the fixed wage / salary. The other career option is self-employment under which one perceives / innovates an idea, organizes production/services by mobilizing resources and finally markets the products and services. Such persons are called Entrepreneurs and they opt for challenging career option.

   Wage-Employment Vs Self-Employment : The two career options available to us differ from each other in terms of nature, scope, orientation, satisfaction, contribution, earning and status. The Wage-employment has always limited scope since it does not necessarily generate resources and can be organized only with the existing usable wealth. Moreover, wage-employment has a strong tendency for self-saturation. Once availed, it blocks the employment opportunities for others for another 10 years assuming average 3 promotions in a 30 year service career. On the other hand, the self-employment contributes towards Gross National Production (GNP) by way of producing consumer items, import substitutes or export goods and services. Further, it has unique characteristics of self-generation. A self-employment activity offers employment to others. Moreover, an enterprise being an economic activity, leads to the emergence of other economic activities such as transport, marketing, communication, etc. This initiates a chain of activities that creates unending opportunities for employment. Indian experiences have further proved that employment generated through self-employment in small industry/business is much cheaper as compared to that obtained through wage-employment in offices or even in large industrial establishments. Also, small enterprises provide learning ground for manufacturing, managerial, technical and problem solving skills. Most of these are generally lacking in masses but are essential for economic progress. Further, small enterprises activate small size savings of middle class individuals for investment in entrepreneurial venture. They also provide an outlet for creative urge among individuals to attain excellence in product design and related innovations. In the ultimate analysis, the lasting solution of the acute problem of unemployment lies in the self-employment and not in the wage-employment.

Career options

	Wage-employment	**Self-employment**
a. Nature	Self-saturating	Self-generating
b. Scope	Limited	Unlimited
c. Orientation	Routine type status quo problem avoiding, Dependent	Creative, Innovative Problem solving Independent (Decision)
d. Satisfaction	Through compliance	Through pursuing own idea and creative urge
e. Contribution	Consumes national wealth	Generates national wealth
f. Earning	Fixed (Subsistence)	Growing and generating surplus
g. Status	Employee	Employer

2. **For increasing national production:** Developing countries have almost a common history - long years of foreign domination, recent emergence as politically free entities and presently engaged in attaining economic freedom. In this process most of them face the problem of increasing national production largely in the areas of import substitute goods, consumer goods and services, and export goods. The first is needed to conserve the outflow of limited national wealth which is needed much more locally for development purposes. But no sooner is the process of all round development initiated in developing countries, the demand for consumer goods and services increases enormously. Such demands are to be met by producing more and more consumer goods and services. This is not enough since no country presently can afford to develop in isolation. On the other hand, for rapid and lasting progress, the developing countries have to trade with others for which they need foreign exchange. One of the major sources of earning foreign exchange is by producing export goods. For all the three, a country needs more and more self-employed entrepreneurs.

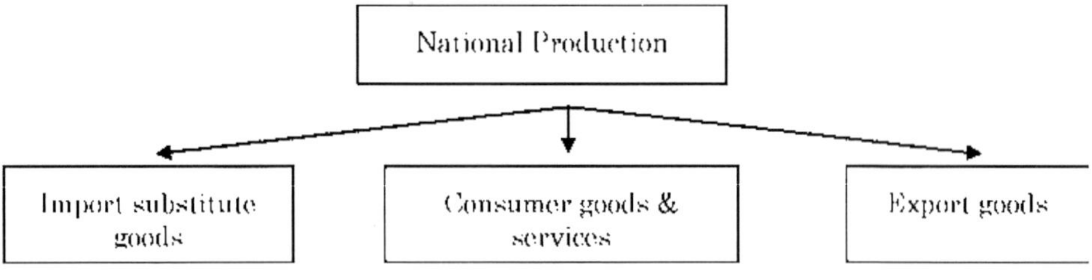

3. **For dispersal of economic power:** Two types of power have dominated the world affairs since its creation - muscle and

economic. In the earlier times, the former was more dominant whereas in the modern society the latter has the upper hand. The phenomenon of economic power is the natural outcome of industrial and business activities. Unless industrial and business activities are planned, economic power is likely to be concentrated only in a few hands. This situation results in many adverse socio-political consequences for developing nations. One way of preventing such consequences may be to disperse industrial and business activities among a large number of youth through a massive movement of developing entrepreneurship / self-employment among the youth of developing nations.

4. **For weakening monopoly:** Monopoly in trade, industry and business holding is the commonest evil in most of the developing countries. In addition to political backing, it largely thrives because of the lack of a sufficient number of entrepreneurs. Planned efforts for encouraging and supporting self-employment among the youth goes a long way in weakening the tendency of monopoly holdings since many self-employed people start manufacturing the goods and services which otherwise remain the prerogative of a selected few.

5. **For balanced area development:** Alongwith the industrial and business activities goes a large number of public benefits, such as roads, transport, communication, health, education, entertainment, etc. Since these industrial and business activities are concentrated in and around selected cities of the country, the development is also limited to these few selected cities. As a result, the country very soon faces the problem of what might be called unbalanced area development. Countless examples can be cited from all developing countries. Till late sixties, in India, more than 50% of the industrial enterprises were located in only six cities. These cities are today much more developed than the rest of the country. Similarly, the city of Bangkok in Thailand, Kuala Lumpur and Penang in Malaysia, Manila in Philippines and Seoul in South Korea are some of the examples of such unbalanced area development. A planned effort to develop self-employment among the local population will ensure the supply of self-employed youth from different parts of the country. Since most of the people, particularly the small entrepreneurs, prefer setting up their units in and around their own native places, more self-employed youth from different parts of the country will ensure the dispersal of industry and business activity throughout the country.

6. **For harnessing youth vigour:** The educated youth are full of vigour with a sharp ability for articulation. The recent

phenomenon of increasing unemployment is largely due to their excessive dependence on wage / salary employment, resulting in decline and wastage of youth vigour. They find themselves without work when they are at the prime of their youth. In the absence of opportunities to meaningfully engage themselves, the youth often divert their vigour to violent and anti-social activities. Many a times they engage in self-destruction. The nation, in short, loses its most valuable resource - the youth vigour. Self-employment offers an excellent opportunity to harness such youth vigour. The task is indeed challenging and creative and it requires extra vigour which is readily available with the youth. In the long run, this also helps in developing self-confidence among the youth, which is usually lacking in developing countries.

Chapter - 3

MOTIVATION AND ENTREPRENEURSHIP DEVELOPMENT

The main function of managers is to achieve predetermined objectives of the organization by organizing the activities of individuals working in the organization. Human factor in production has an important role to play in the organization. The main question for initiating action is how to utilize the availability of human resources efficiently. The efficiency of a person depends upon. (i) The level of ability to do certain work, and (ii) The willingness to do the work. Thus, the factor number one can be acquired by education and training while second factor can be created by motivation.

It is important to understand that the ability to do a work and willingness to work are two different qualities. A qualified person can be appointed on the basis of his ability to do certain work and further can be increased by training. The more important role of managers is how to develop willingness to do the work. To harness their latent capability the manager has to ignite willingness the use of motivation.

Whenever a person does some work there is always a need behind it which motivates him in doing so. This impelling need is called motive. The will to do a work is aroused among the people by making them feel the motive behind their work.

Meaning and definition of motivation

The word motivation is derived from the root *'motive'* which means the latent power in a person which impels him to do a work. In brief, motivation is that process wherein the persons are made to know their needs for the realization of which they will do some work for the interests of the enterprise.

Many scholars have given different definitions on the basis of their experience; some of them are given below.

- Motivation is the act of stimulating someone or oneself to get a desired course of action or to push the right button to get a desired reaction. *- Michael J. Jucius*

- The concept of motivation is mainly psychological. It relates to those forces operating within the individual employee or subordinate which impel him to act or not to act in certain ways.

  - Dalton E. McFarland

- To motivate is to induce people to act in a desired manner.

  - Koontz and O'Donell

- Motivation means a process of stimulating people to action to accomplish desired goals. *- W.G. Scott*

Motivation means movement or motion, an inner state that energizes, activates or moves and directs human behaviour towards goals. It is a need satisfying and goal seeking behaviour.

Importance of motivation

Motivation is an important factor in determining the efficiency of an organization; it develops a desire in the minds of the employees to achieve successfully the objectives of the enterprise. An enterprise may have the best of material, machines and other means of production but all these resources are meaningless so long as they are not utilized by properly motivated people. There was a time when the human resource of production was treated like other non-human resources and was not given any special importance. The old concept has lost all importance in this competitive age. Motivation leads to high enthusiasm among employees. Enthusiastic workers productivity is miraculously increased.

The importance or need of motivation is stressed as;

1. High level of performance.
2. Organizational changes.
3. High employee's turnover and reduction in absenteeism.
4. Good image of organization.
5. Good human relations.
6. Increase in morale.
7. Proper use of human resources.
8. Helpful in achieving goals.

The above review makes it clear that motivation is an important tool in the hands of management for the attainment of objectives of the industry and the organization economically and effectively. Motivation can be used for improving the qualities of subordinates and control on wastages.

Process of motivation

The study of motivation primarily means an attempt to know from where it starts and where it ends. Koontz and O'Donnell have shown the process of motivation as 'Needs-Gap-Satisfaction' chain in the following way:

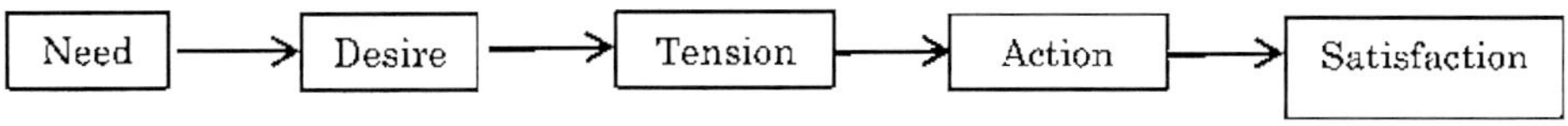

Figure 4 : Process of motivation.

In the first stage a man feels the need of something; a desire is born in his mind. When a need becomes powerful it gets changed into a want which means that he starts feeling that it should be satisfied.

When thoughts come to his mind, tension is created which makes him dissatisfied. The only way to get rid of tension is to do some work. Hence, the situation of action is created. When a man acts to remove the tension created by his needs, he feels satisfaction.

Employees are made to realize their needs through motivation and a feeling to achieve them is born in their mind. The managers make available all the facilities for work performance and, finally, the employees achieve their objective by satisfying their needs. After one need is satisfied, the other is born and the chain continues. It must be made clear that the objective of motivation is not only to satisfy the needs of the employees but to achieve successfully the objectives of the organization.

A state if need is a motive, the action initiated or the drive behavior is known as motivated and attainment of needs and reduction of drive by the achievement of goal is known, as the 'process of motivation'. Motivation in general is an important determinant for entrepreneurial growth and development in society. However, under the broad spectrum of motivation, certain social motives have been found to be significantly related to entrepreneurial behavior such as the need for achievement, power, affiliation, dependency, extension, personal achievement, social achievement, influence etc. Among all these motives, three categories of social motives seem to be extensively explored. These are the need for (a) achievement, (c) power and (c) affiliation.

These categories of social motives differ with each other in terms of satisfaction derived in undertaking activities as well as in terms of thoughts, feelings and actions of the individual.

Entrepreneurial motivation therefore, may be viewed in terms of a set of motives of varying strength such as a high need to achieve, moderate need for power and low affiliation motive. Apart from the motivational profile, other behavioural dimensions like tolerance to ambiguity, problem solving, creativity, etc., adds in building entrepreneurial motivation in a larger perspective. Among these motives, the achievement motive acquires a strategic and important position, and for this reason it is often equated with entrepreneurial motivation which is quite close but is not the same.

People create enterprises in order to grow and fulfil themselves *i.e.,* for self fulfilment. Abraham Maslow has referred to this need as 'self-actualization', a need to maximize one's potential, whatever it may be. It is a desire to become what one is capable of becoming. A person normally does not become an entrepreneur for affiliation or recognition, but to achieve what he thinks he is capable of. This urge to self-actualization is common to all human beings.

Maslow went further and distinguished between 'special talent creativenesses' and 'self-actualizing creativenesses'. While special talent creativeness might be focused on, say, musical or artistic activities, self-actualizing creativeness can just as easily be displayed in innovative and entrepreneurial pursuits. Human beings are naturally creative and innovative. However, all successful innovators have to learn to overcome the emotional, perceptual and cultural barriers, which prevent them from testing and implementing their ideas.

McClelland identified several motivating needs which are basic to entrepreneurship development.

1. Need for achievement
2. Need for independence
3. Need for power

Need for achievement: The need to excel, known as achievement, is a critical factor in the personality of an entrepreneur. People with high need for achievement have desire for success in competition with others, or with a self-imposed standard of excellence. They try to accomplish challenging tasks. Entrepreneurs have been found to be people with a high drive, high activity level and goal orientation. They take external help whenever needed and feel happy on accomplishment of the task.

Need for independence: Studies conducted on entrepreneurs indicate that their need for independence is the prime characteristic that has driven them to start their own business. The entrepreneurs do not conform to routine jobs and practices. They set their own challenging goals and make efforts to achieve these goals. The entrepreneurs do not wait for directions from others and choose their own course of action. They are masters of their own activities and take full responsibility for the outcome of their actions. The independence provides opportunity for trying out new ideas and helping them to achieve their life goals.

Need for power : High need for achievement leads one into launching an enterprise but it may not be adequate to contribute to its success. Once an entrepreneur starts an enterprise and wants to manage it successfully, he/she also needs to influence people, a drive which sells them his/her ideas and leads them in the process of establishing and expanding the organization. Such drive to influence people and to lead them to implement his/her ideas may be called as need for power. It implies controlling the actions and activities of other people. The entrepreneurs, especially in the initial stages, reflect moderate need for power. This helps them to become successful enterprise builders. However, they do not develop emotional bonds with people they work with; and their need for affiliation is low.

Maslow's Hierarchy of Needs

Maslow has given a pyramid of needs depending on importance and urgency of satisfaction of needs. The physiological needs which are very basic to sustaining life are naturally the most important needs.

Physiological needs: Physiological needs are those required to sustain life - air, water, nourishment, sleep, etc. If such needs are not satisfied then one's motivation will arise from the quest to satisfy them. Higher needs such as social needs and esteem needs are not felt until one has met the needs basic to one's bodily functioning.

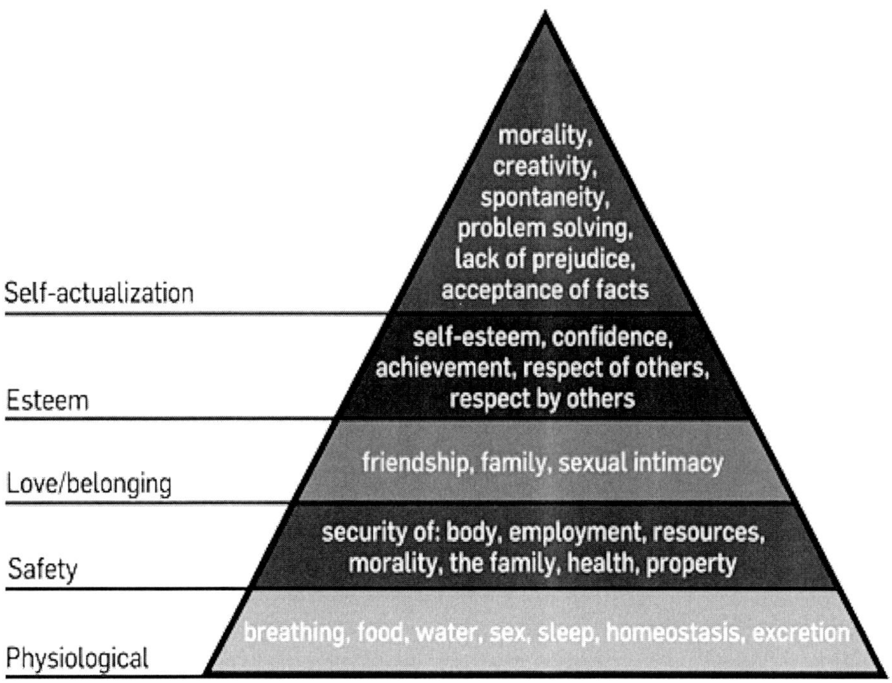

Figure 5 : Maslow's Needs Hierarchy Pyramid.

Safety needs: Once physiological needs are met, individual's attention turns to safety and security needs. Such needs might be fulfilled by - living in a safe area, medical insurance, job security, financial reserves etc. If a person feels that he or she is in harm's way, higher needs will not receive much attention.

Social needs: Once a person has met the lower level physiological and safety needs, higher level needs become important, the first of which are social needs. Social needs are those related to interaction with other people and may include - need for friends, need for belonging, need to give and receive love. These are affiliation needs.

Self-actualization needs: It is the quest of reaching one's full potential as a person. Unlike lower level needs, this need is never fully satisfied as one grows psychologically there are always new opportunities to continue to grow. Self actualized people tend to have needs such as truth, justice, wisdom, meaning etc. Self actualized persons have frequent occurrences of peak experiences, which are energized moments of profound happiness and harmony. Only small percentage of the population reaches the level of self –actualization.

Implications of motivation for management

Managers should see and use opportunities to motivate employees through management style, job design, company events and compensation packages. Perceived needs and desire to satisfy them act as a strong force for arousing motivation.

Physiological needs: Provide lunch breaks, rest breaks, and wages that are sufficient to purchase the essentials of life.

Safety needs: Provide a safe working environment, retirement benefits and job security.

Social needs: Create a sense of community via team based projects and social events.

Esteem needs: Recognize achievements to make employees feel appreciated and valued. Offer job titles that convey the importance of the position.

Self-actualization needs: Provide employees a challenge and the opportunity to reach their full career potential.

However, it has to be borne in mind that not all employees are driven by the same needs. At any time different employees may be motivated by different factors. It is important to understand the needs being pursued by each employee. To motivate an employee, the manager must be able to recognize the needs level at which the employee is operating, and use those needs as levels of motivation.

Implications of motivation for marketing

Advertisement can be used to emphasize either the physiological benefits of a product, such as its nutritional value, or its convenience as a means of satisfying these needs. A food or drink manufacturer marketing a luxury product may however find that, rather than appeal to the physiological need for food and drink, it is preferable to identify his product with satisfaction one of Maslow's higher needs. He may appeal to belonging needs and display his product in a social context. Alternatively, he could appeal to esteem needs and suggest that by serving the product to friends the social status of the consumer will be enhanced.

Motivating factors for entrepreneurship development

The common man thinks that people go into business and become entrepreneurs solely to make money. The desire to earn money is no doubt an important motivating force. But entrepreneurs are not moved by profits alone. Many other factors inspire entrepreneurs to perform their job successfully. Several research studies have been conducted in India to identify the motivating factors that inspire entrepreneurs. Sharma (1987) identified the following internal and external motivating factors.

1. **Educational background:** Education equips an individual for solving problems of life. The educational background helps him in understanding the problems and finding solution to it. Education helps in doing thing in better way.
2. **Occupational experience:** Experiences help an individual by going through the situation. It provides him the learning experience.
3. **Desire to work independently:** This helps an individual in taking independent decisions and gain control over the situation. Success gives him confidence in performing the job in future.
4. **Desire to branch out to manufacturing:** The desire to do something new makes an individual to think differently. He may start a branch or section which is altogether different from the main firm. But mostly a complimentary branch is helpful to the main branch. Out of experience and study one can take such steps.
5. **Family background:** An entrepreneur born in an industrialist family inherits certain social and cultural background which is conducive to his work.
6. **Assistance from government:** There are a large number of schemes to help the entrepreneurs in developing their enterprise. These help motivate him to expand or start new business.

7. **Assistance from financial institutions:** The banks and other financial agencies provide assistance for the enterprise. These institutions give funds as per the project proposal submitted by the entrepreneur, which itself acts as a motivating force.
8. **Availability of technology and raw material:** The new technology is being evolved everyday. Sometimes the raw material is available for the enterprise but there is no one to take advantage of this situation. The entrepreneur should organize the things in such a way that it will develop his enterprise. He is motivated by the opportunities in front of him.
9. **Other factors of production:** There may be many other factors of production like heavy demand for particular product, utilization of excess money earned from contractual estate business, started manufacturing to facilitate trading / distribution business, no chance for further promotion of present enterprise which may motivate an individual to start new enterprise.

The above nine factors are grouped into two major categories internal and external. The first five factors are termed as internal and the last four factors as external. The internal motivating factors like education, occupational experience, family background, the desire to do something independently together make the personality of the entrepreneur. These factors generate an inclination to start entrepreneurial activities.

The internal factors are important but they need to be supported by external factors like favourable governmental policies, financial assistance, technology and raw material and infrastructural facilities. These facilities or assistance are the external motivating factors and serve as a spark in the lightening of the entrepreneurial idea. These factors also provide a momentum to the entrepreneurial activities.

Management of an enterprise

Launching of an enterprise is an important phase. However, one has to remember that efficient management of an enterprise is equally important which is not an easy task. Management involves a number of functions to be played by entrepreneur or a manager. The basics of management of an enterprise are as follows:

Planning: The enterprise involves selecting objectives and strategies, policies, programmes and procedures for achieving them. Planning also includes decision making on production, pricing and marketing of products. Market survey is essential to get an idea of the market. Market survey may be conducted with reference to the availability of raw materials, equipments, marketing and distribution, and consumer behaviour, etc. Each of these functions is critical to the success of any manager and organizations.

Organizing: It involves establishing an intentional structure of roles through determination of activities required to achieve the goals, grouping of activities, delegation of authority and coordination, etc.

Staffing: It demands defining the workforce, utilization of manpower, appointment, promotion and remuneration including downsizing and organizational unity.

Leading: Leading the entrepreneurial unit in terms of addressing the desire, attitude and behaviour of individual and groups amidst challenges towards opportunities.

Controlling: It is the measuring and correcting the activities of staff to assure that events conform to plan. It measures the performance against goals.

Finance: It is most crucial that warrants the personal rapport with the financial institutions to get it materialized. The entrepreneur has to show the required faith and credibility, and strength of the enterprise to get the required credit.

Quality control: It determines the future of one's aspiration climb up. Quality products create wide market, thus making the entrepreneur's task of intervention easy amidst stiff challenges.

Marketing linkage: It should be diligently built up by making the products cost competitive, unique and indispensable before the consumers.

Alternate opportunity: It refers to the next best option that could be explored or employed on the face of parallel enterprise and challenge to one's entrepreneurship venture.

Planning

Planning is a process which involves studying the past and present in order to forecast the future and in the light of that forecast determining the goals to be achieved and what must be done to reach them. Planning is an analytical intellectual activity of deciding the goal and way to reach to it as the most appropriate course of action. Planning is designing a course of action to achieve ends. Thus, planning is rightly defined as, the process of setting goals, developing strategies and outlining tasks and schedules to accomplish the goals.

Planning the enterprise involves selecting objectives and strategies, policies, programmes and procedures for achieving them. Planning also includes decision making on production, pricing and marketing of products. Planning is an integral part of any fruitful activity. The need for planning arises from the complexity and from the importance of the job to be done. It is necessary aspect of any activity. The use of planning is to discover and prepare the way for action that should be taken. It gives meaning and system to action. It prepares the basis for a course of future action.

Starting an enterprise is a big event in the life of an entrepreneur and he should not start it without planning. If he starts the enterprise without planning and proper preparation, then he may meet failure.

- A plan is a trap laid to capture the future.
- Planning is the process by which you determine whether you should attempt the task workout the most effective way of reaching your target and prepare to overcome unexpected difficulties with adequate resources.
- It bridges the gap from where the organization is and where it wants to be. It provides a framework within which a business must operate.
- It is a proactive process that is intended to help individuals; group and organization's to perform effectively to attain objectives.
- However it is seen that people avoid planning because of organizational problems such as poor effort and reward structures, fire-fighting, the get-stuck in culture, non monetary motivation or individual problems such as laziness, lack of commitment, resistance to change, fear of failure and lack of experience.

Importance of planning

- Planning is just one of the vital ingredients needed for success.
- Planning enables you to turn your dreams into reality.
- Planning is an essential stepping stone to business success.
- Failing to plan to win is the same as planning to lose.
- Planning gives road map and a blue print for action.
- It helps directing integrated efforts towards objectives.
- Planning helps to take caution against odds and constraints likely to come up in future.
- Planning helps organizing resources in advance.
- Planning gives an opportunity to develop enterprise in planned manner.
- If you fail to plan, you plan to fail.

Planning is the premier function

Because

- without planning, you cannot organize
- without organization, you cannot direct
- without direction, you cannot control
- without control, you cannot get results
- therefore, for success / results, you need planning

Planning provides

- sense of purpose
- sense of direction
- strategic outlook
- mean for action
- sense of timing
- mean to allocate resources

Functionally, planning provides opportunity to

- determine the environmental impact on the business.
- assess the organization's strengths/weaknesses.
- determine the business opportunities/ threats to business.
- develop strategic plans for the company.
- develop long term/short term plans.
- develop a vision for the organization.
- develop a mission statement for the organization.
- develop business objectives for the organization.
- develop business strategies for the organization.
- develop the action / implementation planning guidelines, which provides the platform for organization and staffing.
- direct the organizational approach.
- develop leadership.
- control the affairs of enterprise.

Steps in planning process

- reviewing the current operational situation.
- conducting the current operation strengths or weaknesses.
- studying the external environmental factors affecting the operation.

- studying the expectations of the operations.
- determine the opportunities for improvements/growth and negatives / constraints.
- based on the above analyses, determine the goals and objectives for the operation for the future period.
- based on the objectives, determine your strategy - how you are going to achieve the objectives.
- based on the strategy, determine the action plan that has to be implemented.
- your action plan will determine the resources required - manpower, finance, materials, etc.
- finally a system to monitor the plan / its progress.

Monitoring

Monitoring can be broadly defined as frequent largely routine-wise collection and analysis of and reporting on information about the performance of the work in a programme or project, comparison of this with the programme or project plans, and connected discussion about and proposals for any corrective action. Thus, monitoring aims at meeting information requirement for current programme and project management.

Generally monitoring function starts after launching an enterprise. However, it is also involved in launching activities. Monitoring involves carefully taking the review of the activities when the work is in progress and seeing, that they (activities) are leading to the objectives as per the plan. If there is little progress or no progress, then the causes of failure need to be identified. The mid term correction is possible by monitoring. One should try to find out the bottlenecks in the programme. It will help in avoiding wastage of efforts, if the things are not moving in the right direction. Monitoring will save the enterprise from disaster if things are not moving in right direction. Continuous checking and supervision of a project is required for implementing it in time and collecting required information for evaluating its progress.

Monitoring of a project is required especially in two types of conditions, viz., (a) Time over runs for completing project and (b) Project cost over runs. Because of time over runs for completing project, the time limit for getting benefits of project are extended and because of project cost over runs, a provision of additional finance is required. Therefore in order to complete project within specified time period and project cost, continuous monitoring of project is required.

Monitoring primarily covers issues of finance and quality pertaining to inputs and outputs as well as actors and time use in implementation. It should usually also encompass some current assessment of direct results (effects) and may cursorily address additional issues, which may possibly be analyzed more thoroughly in some subsequent evaluation. Most monitoring is usually done by agencies responsible for programme or project.

Evaluation and Follow up

Evaluation is the process of determining the extent to which objectives have been attained. The word 'Evaluation' has its origin in the Latin word *'valerie'* which means to be strong or valiant. Its dictionary meaning is determination of the value or estimation of the forces for making a judgment of something. However, the process of evaluation is truly a common happening. We evaluate our activities in everyday life, knowing or unknowingly. We write a piece of information, and we read it and reread it to see whether the expected meaning has been incorporated in it or not. This is evaluation. We draw a picture and then look to it critically whether we are able to draw the picture of an intended animal? For example, we wish to draw the picture of horse and unfortunately we find that it looks like donkey. This is evaluation. Then we may recorrect the picture or take another paper and draw it fresh. Evaluation is an integral part of developmental process and an essential ingredient of agricultural development. In other words, it is the process of determining the extent to which one has been able to attain the objectives.

Evaluation can broadly be defined as a thorough examination, at specific points in time, of programmes or projects or parts of them, usually with emphasis on impacts in addition to efficiency, effectiveness, relevance, replicability and sustainability.

The process of evaluation involves four steps.

1. Making observations and collecting information about an act, activity, programme or enterprise.
2. Comparing the observation with some standard or criteria.
3. Making judgment about whether the performance meets the standard or not.
4. Making decision about corrective measures to be applied to meet the standard.

Generally for developmental programmes, predetermined objectives serve as standard against which performance is measured. It is a process by which the values of an enterprise are ascertained. It is a method of determining how far an activity or programme has progressed and how much further it should be carried to achieve objectives. It is the measurement of programme against the predetermined goals or it is a comparison of situation before and after the programme.

Evaluation means making an appraisal of the strength or worth of an enterprise by collecting and analyzing quantitative and qualitative data about its performance, and pursuing the initial effort by supplementary action.

Follow-up is the action and review of the enterprise from time to time. It is not enough to plan the enterprise; it needs to be brought into practice as decided in the plan and manage it further on sustainable basis.

In follow-up, the plan of work needs to be brought into action, if lacuna is observed, then corrective measures should be taken. If the present activities fail in reaching the objectives, then the enterprise need to be set on sound footing. The activities will have to be corrected to reach the desired objectives. Follow-up will help the entrepreneur in deciding the right course of action.

Follow-up is making an appraisal of the strength or worth of an enterprise by collecting and analyzing quantitative and qualitative data about its performance, and pursuing the initial effort by supplementary action.

Managing Competition

The rivalry between business organizations having similar interest is a common phenomenon. For the business to survive in the face of stiff challenge and to ensure its sustained growth in that environment, the entrepreneur has to adopt certain principles of managing competition. Some of them are –

- Spot early opportunities and utilize them.
- Develop a deeper understanding of the customer – local, national and international.
- Keep track of the competitors.
- Identify current trends which would shape the future.

Chapter - 4

ENTREPRENEURIAL BEHAVIOUR

Behaviour, in simplest terms, means response of an individual to stimulus. However, it has to be born in mind that behaviour does not only mean the observable or visible responses of an individual. In fact non-observable behaviour forms a major part. The observable behaviour is called as overt behaviour and non-observable as covert behaviour. Covert behaviour directs the overt behaviour and in this sense is more significant. Stimulus-response theory is simple form of presentation. However, behaviour is a complex phenomenon. Many mental processes and forces act between stimulus and response. Whether an individual will respond to a stimulus and if he does do so, in what way he will respond depends on a multiplicity of factors. The need, desire, motivation came in between and decide the manner of response. The inner mental processes are complex and they decide the act of an individual in respect of the given stimulus. Not only the internal forces decide the behaviour, but many social, cultural, normative factors and the social milue in which individual placed – all these decide the behaviour. A bundle of pull and push factors act on the individual. We have to look to the entrepreneurial behaviour, which is rather still mere complex, from this point of view.

Entrepreneurial Behaviour

The entrepreneurial behaviour consists of the following qualities in the entrepreneur

1. He copes well or even thrives in uncertainty.
2. He is creative problem solver.
3. He has strong human and organizational skill.
4. He understands the relationship between organizational strategies and environment.

Factors affecting the entrepreneurial behaviour of the farmers

1. **Money:** The entrepreneurs constantly emphasize shortage of money and its effect on business and living. They see it as influencing business investment in three ways;

a. The supply of labour under their control.

b. The expenditure on farm requirement, particularly on commonly used items like seed and other inputs and

c. It prejudiced substantial long term investment because the farm people could not find the means to live till the fruits are to be received.

2. **Labour:** The farmers use less labour than they wish because their supplies of unpaid family labour and of cash to hire labour are limited. Shortage of farm labour is assuming as an acute problem recently.
3. **Size of farm and size of farm business:** The size of farm in hectares is only one of several factors that determine the size of farm business it supports. The farmers face shortage of land as a preventing factor expanding their farm enterprise.
4. **Management:** The farmers work on the farm as well as manage the farm enterprise. In the factory production various jobs like buyer, mechanic, manager, book keeper and salesman are performed by different persons. However, most of the farmers perform all the functions alone. Many of them are not trained in performing these functions.

Inadequate irrigation facilities, natural calamities, uncertainty of production, unfriendly marketing network are few other factors that influence entrepreneurial behaviour of farmers adversely.

Factors affecting entrepreneurship

Entrepreneurship is a human skill, which can be developed. Due to entrepreneurship development, living standards of society can tend to raise, new opportunities of employment can be created and rapid economic and industrial development become possible. Due to the importance of entrepreneurship, it becomes necessary to study factors affecting development of entrepreneurship. The factors affecting entrepreneurship or motivating entrepreneurs can be presented as below:

1. **Individual quality:** Entrepreneur is an individual having specific knowledge, skills and efficiency. Any new enterprise is created by an individual or group of individuals. The creativity of an individual encourages him to establish a new enterprise. Creativity consists of innovation, search and research. Such skills are not shown in all individuals. Personality, social conditions, supporting society, higher education, training and such other factors play important role in developing such skills. Thus characteristics affecting an individual like skills, motives,

attitudes, social-cultural conditions etc. motivate an individual to become an entrepreneur.

2. **Industrial environment:** More suitable the industrial environment in a nation, more rapid development of entrepreneurship is shown. More favorable industrial environment is one, where transportation, communication, electricity, labor, water, raw materials, etc., are easily available. Such industrial environments by and large affect entrepreneur development.
3. **Social environment:** Better and more appropriate the social environment in a nation for entrepreneurship, more rapid development of entrepreneurship is observed. Social system plays vital role in providing certain type of social environment. As an individual is borne and developed in a family and society, social values, ethical standards, family structure, caste and religion, attitudes are part of social environment that affect entrepreneurship development.
4. **Economic environment:** The economy in which enough capital funds are available for establishing an enterprise and market incentives are also available encourages entrepreneurship development. Banking, education, industrial policy, economic policy, Exim policy, interest rate, etc., factors of economic environment affect entrepreneurship development. Thus economic soundness and free economy motivate entrepreneurship development positively.
5. **Technological environment:** Technology is an art of converting the natural resources into goods and services more beneficial to society. Higher the technological development, more the entrepreneurship development is universally accepted. Due to technological development, new product, new production process, new raw materials, new researches are encouraged for modernization. So it can be said that the country, in which technological environment is more suitable, affects remarkably the entrepreneur development. Due to this reason, presently rapid entrepreneurship development has been reported in the countries like Japan, America, and China.
6. **Political environment:** Government also plays important role in entrepreneurship development. If the existing government frames industrial policy supporting industries, it encourages more and more entrepreneurs. Due to globalization, Indian economy has adopted free industrial policy, restrictions on industries have been minimized and MRTP act has been cancelled, which has motivated many entrepreneurs to establish and to develop industries in Indian economy. Thus political environment, having less interference of state and central govt.

and less restrictions on industries, encourage entrepreneurship development.

7. **Incentives:** Incentives are also one of the important factors affecting entrepreneurship. If motivating plans, policies, organizations are developed, it leads to rapid entrepreneurship development. Entrepreneurship development is definitely shown in the economy where there are planned systems, social institutions are established, people are purchasing newly developed product, development oriented programmes are implemented and required facilities and aids are provided to entrepreneurs.
8. **Profit making:** It is the profit that induces the prospective entrepreneurs to get into the business and start new commercial activities. Therefore, profit is a factor which induces the entrepreneurs to organize and utilize the factors of production for development. It does not necessarily mean that the entrepreneur is concerned with profit only. He also satisfies many social needs.

Innovations and entrepreneurs

Entrepreneurs set up ventures, which shift the resources into area of higher productivity and greater yield. Pure entrepreneurs depend upon innovation to set up new ventures. In a fact, entrepreneur is himself an innovator, who introduces something new in the economy. Entrepreneurs are looking for ways to encourage and foster creativity and innovation.

Meaning and definition of innovation

Entrepreneur is an innovator, because he introduces something new in an economy. By innovation, an entrepreneur shifts the resources into area of higher production and greater yield. Innovation plays dominant role in creation of new product.

Innovative behaviour is very important entrepreneurial function. Therefore it is reveal that innovation is the process of doing new things. It implies action, not just conceiving new ideas. Innovation is a process of introducing new product or production process and implementing it in production function.

According to Joseph Schumpeter, innovation means 'doing all the things that are not generally done in the ordinary course of business like new manufacturing process, introduction of a new good or service, locating a new source of raw material, opening a new market, developing new combination of means of production etc.

Peter Drucker states that systematic innovation consists of the purposeful and organized search for changes and in the systematic analysis of the opportunity such changes might offer scope for economic and social innovation.

In short, innovation is a specific tool of entrepreneur by which he exploits changes as opportunity for different business or service. By innovation, an entrepreneur does things in a new and better way.

Creativity and innovation

The terms creativity and innovation are often used to mean the same thing they. But they are not the same, though they are closely related. Both have a unique combination. Creativity is the ability to bring something new into existence and innovation is a process of doing something new. Ideas have little value until they are converted into innovations, new products, services or processes.

Therefore, innovation is the transmission of creative ideas into useful application, but creativity is a prerequisite to innovation.

- Creativity is the ability to bring something new into existence.

  - *Webster's New International Dictionary.*

- Creativity is the ability to see (or to be aware) and to respond.

  - *Erich Fromm*

- Creativity involves the application of a person's mental ability and curiosity to some ideas, with the creation or discovery of something new as a result. - *Hicks and Gullett*

Relationship between creativity and innovation

- Creativity is the ability to bring something new into existence, while innovation is the process of doing new things.
- Creativity is the ability to bring something new into existence, so the emphasis is ability and not on bringing something new as innovation.
- Creativity is pre-requisite to innovation. Innovation transforms creative ideas into useful application. That is why entrepreneur has been viewed as a creative and innovative response to the environment and an ability to recognize, initiate and exploit an economic opportunity.
- Creativity without innovation is aimless, while innovation without creativity is sterile.

The creative process –

Clearly action by itself has no meaning; it is of little value to simply 'do things' without having inspiration and direction. Entrepreneurs need new ideas to pursue and ideas seldom materialize accidentally. Ideas usually evolve through the creative process, where by imaginative people germinate ideas, nurture them and develop them successfully. Various labels have been applied to stages in the creative process, but most social scientists agree on five stages described by David Halt. The five stages are:

a. Idea germination
b. Preparation
c. Incubation
d. Illumination
e. Verification

Relationship between invention and innovation

- Joseph Schumpeter has made a distinction between invention and innovation. Invention is the creation of new products, processes or technologies not previously known to exist. On the other hand, innovation is the transformation of creative ideas into useful applications by combining resources in new or unusual ways to provide value to society for improved products, technology, or services.
- Invention means discovering new methods and new materials and innovation means utilizing or applying invention and discovering higher and better quality goods that give greater satisfaction to the consumer and higher profit to entrepreneurs.
- An inventor produces ideas and an innovator implements them for economic gain.
- An inventor adds to the knowledge of society, while an innovator adds to their satisfaction.
- It is an innovator who commercially exploits invention.

Characteristics of innovation

1. It is a process.
2. It is a continuous process.
3. It has universal application.
4. Invention is pre-requisite to innovation.
5. It gives economic and social gains
6. Creative wealth.

7. Important function of an entrepreneur.
8. It has different forms.
9. Requires specific qualities.

Types of innovation

According Joseph Schumpter, there are five types of innovations as below:

1. Introduction of new product with which consumers are not familiar or introducing a new quality in the existing process.
2. Introduction of new manufacturing process that has not yet been tested and commercially exploited.
3. Locating a new source of raw material or semi-finished material that was not exploited earlier.
4. Opening a new market, where the company products were not sold earlier.
5. Developing a new combination of means of production.

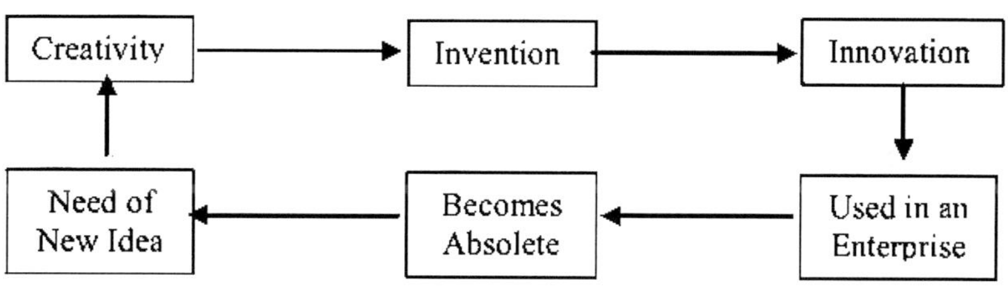

Figure 6: Innovative cycle in an enterprise.

Chapter - 5

GOVERNMENT POLICY AND PROGRAMMES FOR ENTREPRENEURSHIP DEVELOPMENT

Government Policy for SSI

Government's attitude and intension towards Small Scale Industries (SSI) are reflected in the industrial policy resolutions.

Industrial policy resolution 1948

In this industrial policy, government stressed the role of SSIs for balanced industrial growth. It was stated that SSIs are particularly suited for the utilization of local resources and creation of employment opportunities. The primary responsibility for developing small industries by creating infrastructure has been provided to state governments. Central government frames the broad policies and coordinates the efforts of State Governments for the development of SSIs.

Industrial policy resolution 1956

It was stated that besides continuing the policy support to cottage, village and small industries by differential taxation or direct-subsidies, the aim of state policy would be that the development of this sector is integrated with that of large scale industry. The focus was to improve the competitive strength of SSIs. To achieve these 128 items were exclusively reserved for production in SSIs and 166 items were reserved for exclusive purchase by government from this sector.

Industrial policy resolution 1977

It emphasized that whatever can be produced by SSIs must only be so produced. The main thrust of policy was effective promotion of cottage, village and small industries widely dispersed in rural areas and small towns. This thinking specified the following things:

1. 504 items were reserved for exclusive production in the SSIs.

2. The concept of District Industries Centers (DICs) was introduced so that in each district a single agency could meet all the requirements of SSIs under one roof.
3. Technological up gradation was emphasized in traditional sector.
4. Special marketing arrangements through the provision of services such as product standardization, quality control, and market survey were laid down.

Industrial policy resolution 1980

The policy focused on the need of promoting SSIs through integrated industrial development between large and small sectors. Industrially backward districts were identified for faster growth of existing network of SSIs. Following measures were specified in the policy:

1. Investment limit was raised for tiny, small, and ancillary units to Rs. 2 lacs, Rs. 20 lacs, and Rs. 25 lacs, respectively.
2. 'Nucleus plants' in each industrially backward district replaced the 'District Industries Centers.' These were to concentrate on assembling the products of SSIs and to produce inputs needed by large number of small units.
3. Reservation of items and marketing support for small industries was to continue.
4. Availability of credit to growing small scale units was continued.
5. Buffer stocks of critical inputs were to continue.
6. Agricultural base was to strengthen by providing preferential treatment to agro-based industries.
7. An early warning system was to establish to avoid sickness and take appropriate remedial measures.

Industrial policy resolution 1990

Main features of this policy for SSIs are as follows:

1. It raised the investment ceiling in plant and machinery for SSIs.
2. It created central investment subsidy for this sector in rural and backward areas and also assistance was granted to women entrepreneurs for widening the entrepreneurial base.
3. Reservation of items to be produced by SSIs was increased to 836.
4. Small Industries Development Bank of India (SIDBI) was established to ensure adequate flow of credit to SSIs.
5. Stress was reiterated to upgrade technology to improve competitiveness.

6. Special emphasis was laid on training of women and youth under Entrepreneurial Development Programme.
7. Activities of Khadi and Village Industries Commission and Khadi and Village Industries Board were to expand.

Industrial policy resolution 1991

The basic thrust of this policy resolution was to simplify regulations and procedures by delicensing, deregulating, and decontrolling. Its salient features are:

1. SSIs were exempted from licensing for all articles of manufacture.
2. The investment limit for tiny enterprises was raised to Rs. 5 lacs irrespective of location.
3. Equity participation by other industrial undertakings was permitted up to a limit of 24 percent of shareholding in SSIs.
4. Services were to launch to solve the problem of delayed payments to SSIs.
5. Priority was accorded to small and tiny units in allocation of indigenous and raw materials.
6. Market promotion of products was emphasized through co-operatives, public institutions and other marketing agencies and corporations.

Industrial policy resolution 2000

Industrial policy 2000 was actually a comprehensive policy package for SSIs and tiny sectors. Main focus of this policy is as follows:

1. The exemption for excise duty limit raised from 50 lakhs to Rs 1 crore to improve the competitiveness.
2. Credit linked capital subsidy of 12% against loans for technology upgradation was provided in specified industries.
3. The third census of SSIs by the Ministry of SSI was conducted, which also covered sickness and its causes in SSIs.
4. The limit of investment was increased in industry related service and business enterprises from Rs 5 lakhs to Rs 10 lakhs.
5. The scheme of granting Rs 75000 to each small scale enterprise for obtaining ISO 9000 certification was continued till the end of 10th plan.
6. SSIs associations were motivated to develop and operate testing laboratories. One time capital grant of 50% was given on reimbursement basis to each association.

7. The limit of composite loan was increased from Rs. 10 lakhs to Rs 25 lakhs.
8. A group was constituted for streamlining of inspection and repeal of redundant laws and regulations.
9. The coverage of ongoing Integrated Infrastructure Development (IID) was enhanced to cover all areas in the country with 50% reservation for rural areas and 50% earmarking of plots for tiny sector.
10. The family income eligibility limit of Rs. 24000 was enhanced to Rs 40000 per annum under the Prime Minister Rozgar Yozna (PMRY).

Industrial policy resolution 2001-02

This policy emphasizes the following:

1. The investment limit was enhanced from Rs 1 crore to Rs 5 crore for units in hosiery and hand tool sub sectors.
2. The corpus fund set up under the Credit Guarantee Fund Scheme was increased from 125 crore to 200 crore.
3. Credit Guarantee cover was provided against an aggregate credit of Rs 23 crore till December 2001.
4. Fourteen items were de-reserved in June 2001 related to leather goods, shoes and toys.
5. Market Development Assistance Scheme was launched exclusively for SSIs sector.
6. Four UNIDO assisted projects were commissioned during the year under the Cluster Development Programme.

Industrial policy resolution 2003-04

The following are the highlights of this policy:

1. Seventy three items reserved for exclusive manufacture in the SSIs sector were de-reserved in June 2003. These consist of chemical and their products, leather and leather products, laboratory reagents etc.
2. Selective enhancement of investment in plant and machinery was raised from Rs One crore to Rs 5 crore. It was for 13 items in stationary sector and 10 items of drugs and pharmaceuticals sector from June 2003.
3. The investment limit in plant and machinery was raised from Rs one crore to Rs 5 crore in October 2004, in respect of seven items

of sports goods to help to upgrade the technology and enhance competitiveness.

4. The Small and Medium Enterprise (SME) fund of Rs 10000 crore was started by SIDBI since April 2004, with 80% of the lending for SSI units. The interest rate was 2% below the prevailing Prime Lending Rate (PLR) of the SIDBI.
5. The Reserve Bank of India raised the composite loan limit from Rs 50 lakhs to Rs one crore.
6. Promotional Package for small enterprises was initiated.

Industrial policy resolution 2005-06

Industrial policy resolution 2005-06 offered following package for SSIs.

1. The Ministry of Small Scale Industries has identified 180 items for dereservation.
2. Small and Medium Enterprises were recognized in the services sector and were treated on par with SSIs in the manufacturing sector.
3. The corpus of the Credit Guarantee Fund was raised from Rs. 1132 crore in March 2006 to Rs. 2500 crore in five years.
4. Credit Guarantee Trust for Small Industries (CGTSI) was advised to reduce the one time guarantee fee from 2.5 to 1.5 % for all loans.
5. Insurance cover was extended to approximately 30,000 borrowers, identified as chief promoters, under the CGTSI. The sum assured would be Rs. 2,00,000 per beneficiary and the premium will be paid by CGTSI.
6. The emphasis was laid on Cluster Development model not only to promote manufacturing but also to renew industrial towns and build new industrial townships. The model is now being implemented, in nine sectors including Khadi and Village Industries, handlooms, handicrafts, textiles, agricultural products and medicinal plants.
7. From above discussion, it is observed that the Government has recognized importance of SSIs for the economy and its intertie towards promotion of SSIs is reflected in various Industrial Policy Resolutions right from the year 1948.
8. Government is aware of the challenges faced by SSIs and has been trying to improve their competitiveness through various measures. These consist of the following;

a. Tax concessions have been provided to SSIs to promote investment in this sector and also to grant relief to small entrepreneurs.

b. Technological facilities have been increased.

c. In order to facilitate adequate flow of credit and efforts have been done.

d. Measures have also been taken to improve infrastructure facilities and promote marketing of products.

e. To improve access to latest information, automation of the Ministry of SSI Office of DC (SSI), Directorate of Industries and District Industries Centers have been set up.

f. Other initiatives such as Advisory and Monitoring Services, Technology Business Incubators, Suppliers Rating Accreditation Services have been taken up.

Institutional Support to Entrepreneurship Development

Starting a new venture requires various resources and facilities. Small scale enterprises generally find it difficult to have these resources and facilities of their own. Adequate finance is life and blood to start and run an enterprise as it facilitates an entrepreneur to have enough funds to procure land, labour, materials, plant and machinery and so on. Recognizing it, government has come forward to support small entrepreneurs by providing them funds and other needed facilities. Government has recognized the important role of entrepreneurs in the industrial development of the country, especially through the SSIs. The Government of India and state governments provide a number of special facilities and incentives to SSIs. The incentives not only motivate entrepreneurs to set up industries in the small-scale sector, but also strengthen the entrepreneurial base in the economy. The government offers a package of services through its specialized institutions and motivates entrepreneurs to take advantage of the various facilities and establish enterprises and flourish. It is hoped that institutional incentives would play a key role in the promotion of small enterprises and ensure their self-sustained growth. The institutions providing assistance to small scale industries are broadly classified into three categories, namely-

1. National Level Institutions
2. State Level Institutions
3. Fund-Based Institutions

1. National Level Institutions

National Level institutions supporting entrepreneurs are as below.

(a) Small Scale Industries Board (SSIB)

(b) Small Industry Development Organization (SIDO)

(c) National Small Industries Corporation Limited (NSIC)

(d) The Khadi and Village Industries Commission (KVIC)

(e) National Training Institutes like NISIET, NIESBUD and IIE

(a) Small Scale Industries Board (SSIB)

SSI Board is the apex non-statutory advisory body constituted by the Government of India to render advice on all issues pertaining to the SSI sector. It provides a forum to its members for interaction to facilitate co-operation and inter-institutional linkages and to render advice to the Government on various policy matters, for the development of the SSIs. The Board was first constituted in 1954. Its term lasts for two years. The Industrial Minister of the Government of India is the chairman of the SSIB. Its key members are state industry minister, some members of parliament, secretaries of various departments of Government of India, financial institutions, public sector undertakings, industry associations, eminent experts in the field, etc.

(b) Small Industry Development Organization (SIDO)

The Office of the Development Commissioner of Small Scale Industries is also known as the Small Industry Development Organization (SIDO). It is an apex body, established in 1954, for assisting the Ministry in formulating, coordinating, implementing and monitoring policies and programmes for the promotion and development of SSIs. It has over 60 offices and 21 autonomous bodies under its management, including Tool Rooms, Training Institutions and Project-cum-Process Development Centers etc. Functions of such main bodies are as follows:

i. Small Industries Service Institutes (SISIs) are operational one in each state. They provide technical support and consultancy services, conduct entrepreneurship development programmes, and export promotion and liaison activities. Emphasizes is also placed on implementation of programmes on modernization, energy conservation, quality control / upgradation and pollution control for the benefit of entrepreneurs.

ii. Regional Testing Center (RTC) provides Testing facilities for product quality upgradation.

iii. Tool Rooms / Tool Design Institutes (TRs/TDI) assist SSIs in technical upgradation, and provide good quality tooling by designing and producing tools, moulds, jigs & fixtures, components, etc.

iv. Product-cum-Process Development Centers (PPDCs) look into their specific problems and render technical service.

v. Central Footwear Training Institutes (CFTIs) develop footwear designing to promote exports.

vi. Sub-Contract Exchanges for Ancillary Development (SCXs) register and create database of the spare manufacturing/service capacity of SSI, create database of requirements of large / medium units and match the requirements with the spare capacity available with small units and arrange Buyer-Seller Meets, organize vendor exhibitions, seminar, workshops for large-small units coordination, quality upgradation, export promotion, etc. and facilitate flow of data on vendor development.

The main services rendered by SIDO office are:

- Advising the government in policy formulation for the promotion and development of SSIs.
- Providing techno-economic and managerial consultancy, common facilities to small-scale units.
- Providing facilities for technology upgradation, modernization, quality improvement and infrastructure.
- Developing human resources through training and skill upgradation.
- Providing economic information services.
- Maintaining a close liaison with the Central Ministries, Planning Commission, State Governments, Financial Institutions and other organizations concerned with development of SSIs.
- Evolving and coordinating policies and programmes for development of SSIs as ancillaries to large and medium scale industries.
- Monitoring of Prime Minister Rozgar Yojana (PMRY) Scheme.

(c) National Small Industries Corporation (NSIC) Limited

The National Small Industries Corporation Ltd. (NSIC) was set up in 1955 with a view to promoting, aiding and fostering the growth of SSIs in the country with focus on commercial aspects of these functions. NSIC continues to implement its various programmes and projects throughout the country to assist the SSI units. The Corporation has been assisting the sector through the following schemes and activities;

i. ***Composite term loan scheme:*** To promote small-scale sector, NSIC has launched a Composite Term Loan Scheme for the benefit of existing and prospective entrepreneurs to acquire land and building, machinery and equipment and working capital under one roof to the tiny units.

ii. ***Hire purchase scheme:*** Indigenous and imported machinery and equipment on easy financial terms with special focus on women entrepreneurs, weaker sections, handicapped and ex-servicemen and SC / ST entrepreneurs are supplied.

iii. ***Equipment leasing:*** It is done mainly to facilitate SMEs to expand their capacities or diversify and / or upgrade their technology according to the needs of the market.

iv. ***Working capital finance:*** This scheme aims at augmenting working capital of viable and well managed units, on selective basis in case of emergent requirements to enable them to pay off their purchase of consumable stores, spares and production related overheads particularly electricity bills, statutory dues.

v. ***Raw material assistance:*** It facilitates availability of scarce raw material either through the domestic market or by importing.

vi. ***Marketing support programme:*** NSIC has been trying to act as a major agency to bring SMEs closer to various governmental purchasing agencies, with the intention of creating confidence in the purchasing agencies about SMEs, and their capabilities to supply goods and services of requisite quality, economic prices and adherence to agreed delivery schedules.

vii. ***Tender marketing:*** It participates in bulk local / global tender on behalf of Small Scale Industries / Enterprises. It is aimed at assisting SSIs with the ability to manufacture quality products but which lack brand equity and credibility or have limited financial capabilities.

viii. ***Integrated marketing support:*** NSIC has been operating an Integrated Marketing Support Programme in which bills pertaining to supplies made by small scale units to eligible purchasers are discounted by NSIC up to a certain specified limit.

ix. ***Government stores purchase programme:*** The units registered with the corporation for participation in government purchase programme are considered at or with individual purchase organizations and derive all the benefits like free supply of tender forms, exemption from payment of earnest money, security deposits, etc.

x. ***Technology upgradation:*** Excellent technical support is provided to SSIs / SMEs through five NSIC Technical Service

Centers. These centers have been recognized by Council of Scientific and Industrial Research for in house R & D. NSIC has set up a Technology Transfer Center. The latest information is provided to on-line connections and networks of computers on matching technology seekers and technology providers are arranged through the Technology Transfer Center.

xi. ***Software technology parks:*** NSIC has set up a NSIC-STP Complex under Software Technology Parks of India (STPI). Software Technology Parks facilitates small scale units to establish their units for 100% export of software and also act as the major point to activate software exports directly through NSIC. NSIC-STP Complex at Okha, New Delhi is one of such Parks set up by the National Small Industries Corporation under the Software Technology Parks of India to promote small entrepreneurs in software development. NSIC-STP provides high speed better communication facilities through VSNL / SATCOM networks, built-up office space and uninterrupted power supply, back-up power through DG sets, a modern business centre and other administrative support.

xii. ***Exports:*** NSIC is providing a complete package of export assistance, testing facilities, pre-shipment credit facility, export incentives etc. apart from exposure to the products of SSEs in trade fairs, buyer and seller meets etc. The Corporation has been endeavoring to increase share of Indian industries in purchases to United Nations Organization, it being the largest single buyer in the world.

(d) The Khadi and Village Industries Commission (KVIC)

The Khadi and Village Industries Commission (KVIC) is a statutory body created by an Act of Parliament in April 1957. The KVIC is supposed to do the planning, promotion, organization and implementation of programmes for the development of khadi and other village industries in the rural areas in coordination with other agencies engaged in rural development wherever necessary.

Training Institutes

There are three national level training institutes. These are:

1. National Institute of Small Industry Extension Training (NISIET), Hyderabad, which undertakes operations ranging from training, consultancy, research and education, to extension and information services.
2. National Institute for Entrepreneurship and Small Business Development (NIESBUD), New Delhi, which conducts national

and international level training programmes in different fields and disciplines.

3. Indian Institute of Entrepreneurship (IIE), Guwahati was established to act as a channel for entrepreneurship development with its focus on the North East.

2. State Level Institutions

State level institutions supporting entrepreneurs are as below.

a. State Small Industrial Development Corporations (SSIDC)

b. State Directorate of Industries (SDIs)

c. District Industries Centers (DICs)

(a) State Small Industrial Development Corporations (SSIDCs)

Incorporated under the companies Act 1956, SSIDC's were set up in different states as wholly owned companies for promoting industrial development in their respective states. The main functions of SSIDC's are as follows;

1. Providing term finance to all small, medium and large industrial enterprises set up in state.
2. Underwriting and directly subscribing to shares, and debentures of industrial enterprises being set up in the state.
3. Preparing feasibility studies, conducting market surveys and motivating private entrepreneurs to set up their industrial ventures in the state.
4. Collaborating with private entrepreneurs to set up industrial ventures in joint and assisted sector.
5. Implementing scheme of Industrial Development Bank of India (IDBI) of seed capital in the state.

(b) State Directorate of Industries (SDIs)

Under the constitution of India promotion and development of SSIs is a state subject. Therefore, the primary responsibility for implementation of policies and programmes of assistance rests with the Directorate of Industries in each state. It acts under the overall guidance of SIDO and concerned central institutions. It performs both regulatory and developmental functions. It functions through a network of district industries offices, industries offices and extension offices at district sub-division and block level respectively.

The main functions of Directorate of Industries are as follows:

1. Registration of small scale units

2. Providing financial assistance
3. Distributing scare and indigenous raw materials to industrial units
4. Granting essentiality certificates for import of raw material
5. Establishing industrial estates and industrial co-operatives
6. Developing industrial infrastructure
7. Undertaking industrial surveys and collecting information
8. Arranging concessions and incentives for industries.
9. Overall administration of village and small scale industries.
10. Maintaining liaison with other agencies for industrial development.

(c) District Industries Centers (DICs)

The District Industries Centers programme was launched on May 8, 1978 for effective promotion of cottage and small-scale industries widely dispersed in rural areas and small towns. These centers are the focal points providing all the services and support required by small scale and village entrepreneurs under one roof. These serve as integrated administrative framework at the district level for industrial development. The main functions of DIC's are as follows:

1. It conducts surveys to know industrial potential of a district keeping in view the availability of raw material, human skills, infrastructure, demand, etc.
2. It prepares an action plan for industrial development.
3. It appraises the various investment proposals received from entrepreneurs.
4. It guides and assists entrepreneurs in buying appropriate machinery and equipment and raw material.
5. It suggests appropriate marketing strategies to entrepreneurs.
6. It maintains links with research and development institutions for upgradation of technology, quality improvement, industrial training etc.
7. It conducts artisans training programmes.
8. It has been assigned operational responsibility for special schemes to provide self-employment to educated unemployed youth.

3. Fund-Based Institutions

Fund-Based institutions supporting entrepreneurs are as below:

a. Small Industries Development Bank of India (SIDBI)

b. Commercial Banks

c. State Financial Corporations (SFCs)

(a) Small Industries Development Bank of India (SIDBI)

The SIDBI was established in 1990 as the apex refinance bank. The SIDBI is operating different programmes and schemes through five regional offices and 33 branch offices. The financial assistance of SIDBI to the small-scale sector is channelized through the two routes - direct and indirect.

i. Direct assistance:

- The loans are available for new ventures, diversification technology upgradation, modernization and expansion of well-run small scale enterprises. Assistance is also available for private sector.
- Small scale sector is eligible for maximum debt-equity ratio of 3:1
- Foreign currency loan for import of equipment are also available to export oriented small scale enterprises.
- SIDBI also provides venture capital assistance to the entrepreneurs for their innovative ventures if they have a sound management team, long term competitive advantage and a potential for above average profitability leading to attractive return on investment.

ii. Indirect assistance

- SIDBI's financial assistance to small sector is primarily channelized through the existing credit delivery system, which consists of state level institutions, rural and commercial banks.
- SIDBI provides refinance to and discounts bills of Primarily Lending Institutions (PLI).
- The assistance is available for - Marketing of SSI product, setting up of new ventures, availability of working capital, expansion, modernization, human resource development, diversification of existing units for all activities

New Initiatives of SIDBI

1. Two Subsidiaries *viz.* SIDBI Venture Capital Limited and SIDBI Trustee Company Limited formed to oversee Venture Capital.

2. Technology Bureau for Small Enterprise formed to oversee Technology Transfer, Match making Services, Finance Syndication and facilitating Joint Ventures.
3. SIDBI Foundation for Micro Credit has been launched to provide financial assistance to the poor and to meet emerging needs of the micro finance sector especially in rural areas.

(b) Commercial Banks

Credit requirement of SSIs is basically of two types - long term loans and working capital. Commercial banks with their extensive network of branches operating nationwide are primary channel for working capital requirement. Banks are required to compulsory ensure mat defined percentage (currently 40%) of their overall lending is made to priority sectors as classified by RBI. These sectors include agriculture, small industries, export etc. The inclusion of small industries in this list makes them eligible for this earmarked credit. With the liberalization of the Indian economy, greater emphasis was placed on meeting the credit needs of SSIs. This was manifest through the following initiatives taken by RBI.

- Credit for tiny sector has been earmarked within overall lending to small industries. In order to ensure that credit is available to all segments of SSI sector, RBI has issued instructions that out of the funds normally available to SSI sector, 40% be given to units with investment in plant and machinery upto Rs.5 lakhs, 20% for units with investment between Rs.5 lakhs to Rs.25 lakhs and remaining 40% for other units.
- Public sector banks have been advised to operationalize more specialized SSI branches at centers where there is a potential for financing many SSI borrowers. As on March 2002, 391 specialized SSI branches are working in the country.
- Single window scheme was extended to all districts to meet the financial requirements (both term loan and working capital) of SSIs.
- Laghu Udyami Credit Card (LUCC) Scheme was launched by public sector banks for providing simplified and borrowed friendly credit facilities to SSI, tiny enterprises, retail traders and artisans.
- Composite loan limit was enhanced to Rs. 50 lakhs from Rs. 25 lakhs.
- Limit on collateral free loans was increased to Rs. 25 lakhs in deserving cases.

(c) State Financial Corporations (SFCs)

The Industrial Financial Corporation of India (IFCI) set up in 1948 used to provide financial assistance to only large-sized industrial undertakings. In order to cater to the financial requirements of a large number of small-scale units, the State Financial Corporation Act was passed. State Financial Corporation Act 1951 was brought into force to enable all the state governments to set up State Financial Corporations as regional development banks. The first SFC was set up in Punjab in 1953. Today there are 18 SFCs in the country, which exist almost in every state. A SFC has a board of directors, a managing director and an executive committee. A SFC can open its office at different places within the state. Presently following assistance is provided to small scale and medium scale undertakings.

1. Providing long-term finance to industrial enterprises having sole proprietary, partnership, company and co-operative society form of business organization.
2. Subscribing equity and debentures of industrial enterprises.
3. Providing financial assistance to small and medium enterprises engaged in service sector.
4. Provide working capital loans and meeting various short-term needs of their clients.

Thus, the Government of India as well as state governments have developed a definite policy and strategy for promoting enterprises and small scale industries. The entrepreneurs have a good scope to launch variety of enterprises with support of government and various promoting institutions.

Entrepreneurship Development Programmes

The entrepreneurship development is a key to achieve overall economic development through higher level of industrial activity. Every individual has qualities of an entrepreneur. If some or other qualities are developed as an entrepreneur, such individuals can become successful entrepreneurs. Entrepreneurship development plays very important role in this direction. The economic and industrial developments of developing countries like India are only due to entrepreneurship development. In India, rapid growth of small scale sector is mainly due to the entrepreneurship development. Entrepreneurship development programme focuses on identifying entrepreneur qualities of an individual, providing training required for entrepreneur programme, preparing related project profiles, preparing entrepreneurs by providing aid with regard to finance, production, technology, marketing, management, infrastructure facilities etc. and trying to solve problems and difficulties of entrepreneurs by showing remedial measures. Therefore, it is in this context that an increasingly important role has been assigned to entrepreneurship development programme for promotion of an entrepreneur.

Meaning and Definition

A programme designed to help a person in strengthening and fulfilling his entrepreneurial motives and in acquiring skills and capabilities necessary for playing his entrepreneurial role effectively is called entrepreneurship development programme. It is necessary to promote his understanding of motives, motivation pattern, their impact on behaviour and entrepreneurial value.

Entrepreneurship Development Programmes may be defined as an action plan to create an entrepreneur with achievement motivation, who can develop business, can be strategic decisions, can cope successfully with the internal and external environment and can fear the risk of investment. It is an entrepreneurial training programme.

In short, Entrepreneurship Development Programme is a continuous process of training and motivating entrepreneurs to set up profitable enterprises.

Entrepreneurship Development Programme is not meant only a training programme, it is a process;

1. To increase motivational skills and knowledge of proposed entrepreneur.
2. To create and improve entrepreneurial behaviour in day to day activities of entrepreneur and
3. To help in developing own venture after implementing entrepreneurship.

Entrepreneurship Development Programme is primarily meant for developing those first generation entrepreneurs, who on their own cannot become successful entrepreneurs. Various institutions, Government's concerned department, NGOs, etc., established for entrepreneurship development plan such programmes train entrepreneurs for specific objectives.

The Entrepreneurship Development Programmes embrace all aspects of business creation, risk-taking, management and the initiative for the socio-economic development and creation of a business culture among the community. These programmes focus to create entrepreneurial qualities among people who will be able to develop business activities, i.e. set up new business and will be ready or willing to fear the risk involved therein. The emphasis is more on the financial and business viability of the project that can create employment opportunities. However, in low income subsistence level communities, the Entrepreneurship Development Programmes focus to create self employment and alleviate poverty. Entrepreneurship Development Programme generates motivation and develops entrepreneurial competencies among the young prospective entrepreneurs.

Development of entrepreneurship incorporates four basic issues, *viz.*;

1. The availability of material resources,
2. The selection of real entrepreneurs,
3. The formation of industrial units and
4. Policy formulation for the development of the region.

All these issues are closely interrelated. Given the resources and the entrepreneurs expected to exploit them, the focal issues that remains is that of the type of the industrial unit, particularly because it affects the proper utilization of raw material and the marketing of the product.

Entrepreneurial development is an organized and systemic development. It is considered as a tool of industrialization and a solution to unemployment problem. One trained entrepreneur can guide others on how to start their own enterprise and approach various institutions. In fact trained entrepreneurs become source of developing industry and economic progress.

An entrepreneur development programme is based on the belief that individuals can be developed. Their outlook can be changed and their ideas can be converted into action through an organized and systematic programme. Entrepreneurship Development Programme is not merely a training programme, it is a process of enhancing the motivation, knowledge and skills of the potential entrepreneurs, arousing and reforming the entrepreneurial behaviour in their day to day activities, assisting them develop their own ventures or enterprise as a sequel to entrepreneurial action.

Objectives of Entrepreneurship Development Programme

The main objectives of an Entrepreneurship Development Programme in India are as follows;

1. To prepare the entrepreneur to take strategic decisions.
2. To let the entrepreneur set or reset the objectives of his business enterprise and work individually and with his team for the accomplishment of such objectives.
3. To develop a broad vision and provide visionary leadership to his enterprise.
4. To prepare the entrepreneur to hear the unexpected business risks for a long time after training.

5. To instill in his mind the values like integrity, honesty and legal compliance.
6. To make entrepreneur subscribe to industrial democracy and employment welfare.
7. To develop and strengthen among the young persons the entrepreneurial competencies or characteristics, especially achievement motivation, hard work, perseverance and optimistic thinking.
8. To develop in entrepreneurs the art of communication and coordination.
9. To enable the entrepreneur to build an effective organization to achieve enterprise goals.
10. To make the entrepreneur analyze the relevant business environment and understand the procedures and laws involved in establishing small scale enterprise.
11. To teach the entrepreneurs about the sources of financial and managerial assistance and help available for small scale business units.

Importance of Entrepreneurship Development Programme

With reference to India, the importance of Entrepreneurship Development Programme can be discussed as below. Entrepreneurship development may create impact as mentioned here below.

1. ***Rapid industrial development:*** Entrepreneurs and entrepreneurship are base for industrial and economic development in a given area. Entrepreneurship Development Programme makes entrepreneurs for creation of entrepreneurship environment. In India, in every five year more and more amount is allotted for Entrepreneurship Development. After freedom, due to Entrepreneurship Development Programme, India has shown rapid industrial and economic development. Because of Entrepreneurship Development Programme, now-a-day entrepreneurs are on increasing in engineering, chemical, electrical, agricultural industries.
2. ***Reduction in regional industrial imbalance:*** Qualities of entrepreneurship can be developed in industrially backward zone, by arranging Entrepreneurship Development Programme. U.P., Bihar, M.P. and such other states have not shown rapid industrial development. Hence for balanced industrial development, Indian government has designed incentive plans as a part of the industrial policy. In industrial backward zone, Entrepreneurship Development Programme arranges various

training programmes and makes individuals aware about government incentives as well train them for entrepreneurship development according to opportunities available. As a result, such states can also develop in parallel with developed states. Thus Entrepreneurship Development Programme helps in removing or reducing regional industrial imbalance.

3. ***Creation of suitable industrial environment for rapid economic development:*** Before arranging Entrepreneurship Development Programme, opportunities available in different zones are identified and accordingly by designing entrepreneurship development programme, entrepreneurship environment is created for creation of industrial and entrepreneurship environment. Various institutions such EDI, IDBI, IFCI, SFC, DIC, EXIM Bank, etc., give their active contribution. EDI's various courses, innovation centers, entrepreneurship environment and support system, strategic international programme, various publications etc. have environment in the state and country. Entrepreneurship Development Programmes are arranged in rural, semi-urban and urban areas. Therefore, in every area, entrepreneurship awareness is spread. Entrepreneurship development programme plays dominant role in creating the most suitable industrial environment for rapid economic development.

4. ***Optimum utilization of available local resources:*** If available resources are utilized for production purpose, new opportunities for local entrepreneurship development and employment generation can be created which lead to rapid industrial and economic development. Moreover, because of creation of employment opportunities, migration of people towards urban areas can also be checked.

5. ***Reduction in unemployment and poverty problems:*** Since our first five year plan, we are facing unemployment and poverty problems. Government has designed various porgrammes like rural development programme, PMEY, etc., in this direction. But Entrepreneurship Development Programme can play very important role in this direction. By arranging Entrepreneurship Development Porgrammes for educated as well as uneducated unemployed, they can be trained for self-employment and entrepreneurs, self-employment can also create employment opportunities for others in future. As a result, entrepreneurship development programme helps in reducing not only unemployment but also poverty problem by creating self-employment opportunities.

6. ***Protection of entrepreneurs from industrial monopoly:*** Some large industrial units have created monopoly conditions in their zones. Thus government has been forced to reserve some goods and services for cottage and SSIs. By Entrepreneurship Development Programme, entrepreneur can be trained and prepared for such reserve sectors and large scale sectors cannot enter in such reserve sectors. Thus due to such decentralization, small scale entrepreneurs can be protected from industrial monopoly of large scale industries.
7. ***Generation of new sources of income for government:*** For management of Entrepreneurship Development Programme, state and central government have to allot a large amount. But government also gets income by levying various taxes on venture of new entrepreneurs. Some export oriented industrial units help government in earning foreign exchange.
8. ***Progress and expansion of social wealth:*** Government's special schemes for generation of rural wealth have been reached to rural areas generation especially through Entrepreneurship Development Programmes. EDP helps to mobilize unorganized and hidden youth of various areas towards planned industrialization. Thus, EDP makes unorganized and rural area's community aware about various government incentives as well as motivates them to be entrepreneurs for taking benefits of these incentives. This improves social standards and as a result helps in progress and expansion of social wealth up to rural areas.
9. ***Development of socialistic philosophy:*** Our economy influence both capitalism as well as socialism. Because of establishment of Entrepreneurship Development Programme institutions, philosophy of socialism has been spread at various levels through utilization of unused local resources, establishment of tiny, cottage, small and large scale industries. Entrepreneurs have accepted concept of social responsibility. In this way, EDP helps in extending the philosophy of socialism.
10. ***Vehicle for social and economic revolution:*** EDP is designed by considering various classes of the society. In rural, semi-urban and urban areas, with a view to direct women towards self employment and freedom. Special EDPs are designed for women, ST, SC as well as ex-service man and physically handicaps i.e. to say that EDPs have played an important role for bringing poor, exploited and weaker class of our society into economic development. Thus EDP is considered a vehicle for bringing social and economic revolution.

11. ***Development of entrepreneurial opportunities:*** EDP is designed and implemented to develop entrepreneurial opportunities for potential entrepreneurs as well as to upgrade managerial skills for the existing entrepreneurs.

Chapter - 6

SETTING UP A SMALL ENTERPRISE

Launching of an enterprise even a small enterprise is not an easy task. Every enterprise involves a risk and an entrepreneur passes through a complex decision making process while launching an enterprise. The steps in setting up a small industrial enterprise shall be:

1. Deciding to go into business

This is the most crucial decision an individual has to take, particularly while shunning wage-employment and opting for self-employment / entrepreneurship.

2. Analyzing strengths / weaknesses

Having decided to become an entrepreneur the young person has to analyze his / her strengths / weaknesses. This will enable him to know what type and size of business would be most suitable to him. This will vary from person to person.

3. Product selection

The next step is to decide what business to venture into, the product or range of products that shall be taken up for manufacture and in what quantity. The level of activity will help in deciding size of business and form of ownership. One could generate a number of project ideas through environmental scanning, short list a few items, closely examine each one of these and go on to a final product.

4. Market survey

It is easy to manufacture an item but difficult to sell. So it is prudent to survey the market before embarking up on production and ascertain the product chosen is in demand, changes in product design required, determining demand-supply gap, extent of competition, your potential share of the market, pricing and distribution policy, etc.

5. Form of ownership

A firm can be constituted as proprietorship, partnership, limited company (Public or Private) co-operative society, etc. This will depend upon the type, purpose and size of your business. One may also decide on the form of ownership based on resources in hand or from the point of saving on taxes.

6. Location

The next step will be to decide on the place where the unit is to be located. Will it be hired or owned? The size of plot and covered area and the exact site will have to be decided.

7. Technology

To manufacture any one item various processes are available. Information on all these available technologies should be collected and the best one identified. This will be useful in determining the machinery and equipment to be installed.

8. Machinery and equipment

Having chosen the technology, the machinery and equipments required for manufacturing the chosen products have to be decided, suppliers identified and then costs estimated. One may have to plan well in advance for machinery and equipment especially if it has to be procured from outside the town, state or country that is, have to be imported.

9. Project report

The economic viability and technical feasibility of the product selected has to be established through a project report. A project report that may now be prepared will be helpful in formulating the financial, production, marketing and management plans. It will also be useful in obtaining finance, shed, power, registration, raw material quotas, etc.

10. Finance

Money is no problem for starting a small scale industry. But an entrepreneur has to take certain steps and follow specified procedures to obtain it. A number of financial agencies will give loan on concessional terms. Under TRYSEM and SEEUY Scheme entrepreneurs are also eligible for subsidies which obviate the need for margin money.

11. Power connection

The site chosen should either have adequate power connection or this should be arranged now.

12. Installation of machinery

Having arranged finance, work shed, power etc. the next step is to procure the machinery and begin its installation.

13. Recruitment of manpower

Once machines are installed manpower will be required to run them. So the quantum and type (skilled, semi-skilled, unskilled, administrative, etc.) of labour have to be determined. Sources of getting desired labour identified and labour / staff recruited. Possibly the labour has to be trained either at the entrepreneur's own premises or in a training establishment.

14. Raw material

The labour will be required raw materials to work upon the installed machines. The raw materials required may be available indigenously or may have to be imported. Government agencies can assist if raw materials required are scarce or imported.

15. Production

The unit established should have an organizational step up. That is, the structure of the manpower proposed to be employed must be determined. This ensure smooth and effective running of the unit. There should not be any wastage of manpower, material or machine capacity installed. If items produced are exported then the product and its packaging must be attractive.

Production of the proposed items should be taken up in two stages.

(a) Trial production

(b) Commercial production

Trial production will help knowing problems confronted in production and test marketing of the product. This will reduce chance of losses in the eventuality of mistakes in project conception. Only after successfully launching the product at test market stage commercial production should be commenced.

16. Marketing

Having manufactured the product, the stage comes to sell it. This is called marketing. Various aspects like way to reach the customer, distribution channels, commission structure, pricing, advertising / publicity etc. would have already been decided upon at market survey / project formulation stage.

Like production, marketing should also be attempted in two stages:

(a) Test marketing

(b) Commercial marketing

Test marketing will save the enterprise from going into disrepute just in case product launched is not well accepted in the market. It will also assist in carrying out modifications, additions in design, characteristics, and features of the product as per the preference of consumer. Having successfully test marketed the product, commercial marketing can be undertaken.

17. Quality control

Before marketing the product, quality certification like ISO, BIS (earlier ISI), 'Q' Mark, Agmark, Del-in, etc., depending upon the product should be obtained. If there is no quality standards specified for the product the entrepreneur should evolve his/her own quality control parameters. Quality, after all, ensures long-term success by establishing reputation and credibility of the enterprise.

Phases of Entrepreneurial Venture

Phase 1: Preparation: Preparation deals with a conscious search for the solutions, seeking information about the problems which arise during the entrepreneurial process.

Phase 2: Incubation: It is the stage of subconscious assimilation of information, where it is allowed to ferment. The process of mental fermentation allows the entrepreneur to collect and assimilate further relevant information and develop clarity of thinking.

Phase 3: Idea generation: Various ideas and solutions are generated at this stage. The approaches to problem solving relating to the new venture are tested in real life situation, using the individual's researches, previous experiences, insights and risks.

This phase also stimulates the decision making process, and helps the entrepreneur to identify a solution relevant to his / her own situation.

Phase 4: Identification of enterprise: This may come from unsatisfied need of an entrepreneur, as well as unsatisfied personal need. This is reflected by the alertness of the entrepreneur to identify the business opportunity.

The most important task in this phase is to identify a solution which is likely to be successful and profitable in the long run.

Phase 5: Initiation / creation of enterprise: The entrepreneur needs to create an organization for transforming the concept into marketable product by utilizing and combining physical and other resources. The entrepreneur interacts with the environment to transform ideas into reality.

Phase 6: Nurturing of enterprise: Organization translates the business concept into marketable products or services, and offers in terms of sales, profitability etc. The feedback is measured in terms of efficiency, competitiveness, effectiveness, innovativeness, flexibility etc.

Phase 7: Conducting transition phase: Entrepreneurial process does not end with achieving stability and success. The organization shall require different managerial style as the environment changes and competition gradually builds up more and more.

Chapter 7

SELECTION OF PRODUCTS

Product selection as mentioned earlier is one of the steps in setting up a small enterprise. However, one has to remember that it is the hardest and the single most important decision takes in this whole business of setting up a small scale enterprise. As the old saying goes 'Well begun is half done', and if your decision to select a product is right, basically your enterprise will do well, if not, there are likely to be troubles ahead. One of the reasons why many small scale units become 'sick' is that no serious thought is given by the entrepreneur before starting the unit on whether there is reasonable chance of succeeding in the product selected. Being this the crucial step, it is discussed at length there.

Even after having taken a correct decision, all sorts of things can happen and upset plans, but in such cases, an entrepreneur can retrieve the situation, provided his product is intrinsically sound. There could be factors beyond control and one must be prepared to face them.

In making your mind what to make, the first thing is to analyze yourself - What is your interest? What are your basic skills? In what field you are good? All of us have strong points and weak ones and in choosing your product you must capitalize on your strong ones. Don't let your weaknesses take over

According to Sharma (1998), the exercise known as 'Micro-screening' may help entrepreneur select a right kind of product. The output of Micro-screening consists in having a group or individual participants identify 10 potentially viable project ideas that the group or individual participant thinks feasible for undertaking in a specific locality (stated by the facilitators, by the location of the training, or by the origin of the participants) from the number of ideas that have been generated as a result of the brainstorming session. The 10 identified potentially viable project ideas are put in a flip chart. A brief presentation by the group representatives or individual participants may be called for by the facilitator.

No individual participant should be forced to present his project ideas as he/ she may reason for their confidentiality. This is specially true where the individual participants stick to their original projects as mentioned by them in their application forms or during the interviews.

The participants should, however, be open to the facilitators to receive guidance and advice in the development of their business plans.

After the presentation by the participants, the facilitator explains that the 10 potentially viable project ideas will undergo a finer screening test that is the Micro Screening. He then proceeds to explain the concept. He uses a ready-made Micro Screen chart containing various columns with the following headings:

PJ	MKT	RM	TEC	SKL	GOP	SFT	El	RE	PFT	C/B	TTL	CSF
1	2	3	4	5	6	7	8	9	10	11	12	13
1												
2												
3												
4												
5												
6												

Column –

1. Project (PJ)
2. Availability of Market (MKT)
3. Availability of Raw Materials (RM)
4. Availability of Technology (TEC)
5. Availability of Skills (SKL)
6. Government Priority (GOP)
7. Strategic Fit (SFT)
8. Ease of Implementation (El)
9. Risk Exposure (RE)
10. Profitability (PFT)
11. Cost/Benefit (C/B)
12. Total (TTL)
13. Critical Success Factor (CSF)

The 10 factors (starting with column 2 to column 11) will be rated according to the participant's or group's perception, availability of information, or experience. In the following manner - 1. Poor 2. Fair 3. Satisfactory 4. Very satisfactory 5. Excellent

The potential total score, a project idea can get is 50 points, that is 5 for excellent multiplied by 10 factors.

The facilitator must explain the meaning and significance of the 10 factors using a specific project as an example. The following is a guide;

1. Availability of market

If there is no assurance of an adequate market, there is no sense in going to business. The market must be large enough to enable the entrepreneur to capture a market share and in the process, make an attractive profit from the transaction after deducting his costs from sales. Indication of availability of market includes, among others:

a. Existing demand is not at present adequately served by existing suppliers;
b. Existing demand is presently served by imports;
c. Existing demand is presently not served at all;
d. The project's product has significant uniqueness or unique selling features – USF (in the case of a product) or unique selling proposition - USP (in the case of a service) such as more desirable features, better quality, more durable, better taste, superior after sales service, free delivery, etc.
e. Supply of the product/service is not reliable;
f. Demand for the product/service is expected to increase significantly or substantially in near future;
g. Supply of the product is presently served through smuggling.

2. Availability of raw material

Availability of raw materials is indicated by the following considerations:

a. Raw materials are available in adequate quantity locally;
b. There is reliability of supply whether local or imported sources;
c. Seasonality, perishability, quality and variability of raw materials have been considered and found to be satisfactory;
d. Price of raw materials is reasonable;
e. Increase in the price of raw materials in the future is perceived to be reasonable and predictable.

3. Availability of technology

Availability of technology can be evaluated in terms of the following indicators:

a. The technology or technologies to be used have been proven;
b. Reasonably priced technologies are available to produce the product;
c. Technology is appropriate for the level of production, level of investment and desired product quality;
d. The project will not suffer from technology obsolescence which will render the project non-viable;

4. Availability of skills

Availability of skills can be gauged by the following factors:

a. Different skills (conceptual, managerial, technical and manual) needed by the project are available;

b. Supply of skills is relatively steady and stable so as not to jeopardize the project in case of sudden or unforeseen labour changes, unusual turnover, or unexpected problems;

c. Cost of labour is projected to be fairly steady and predictable.

5. Government priority

Government priority is indicated by the following considerations :

a. The project is listed under the government's list of priorities for promotion or investment;

b. The project receives government incentives whether fiscal (e.g., tax exemption or-reduction, tariff protection, import privileges, etc.), monetary (priority lending status, reduced interest rate), or other support assistance (e.g., marketing, technical or consultancy services);

c. The project falls under the government's priorities of import substitution, export promotion, employment generation, rural industrialization programs, technology development / transfer, etc.

6. Strategic fit

The strategic fit of the project under consideration is indicated by the following criteria:

a. Proposed project fits well within the competence and expertise of the entrepreneur or key staff;

b. Proposed project fits well within the existing product line, technology, marketing set up, production system, facilities and resources of the entrepreneur or the firm;

c. Project complements and enhances the existing set up, viability or growth of the firm through a positive synergy.

7. Ease of implementation

Ease of implementation can be measured by the following criteria :

a. Project can easily be implemented because the inputs are readily available;

b. Project can be implemented within a short gestation period or reasonable preparatory period (e,g., 3 months to one year);

c. Project can start operation within one year from the completion of the training;

d. An unforeseen difficulty can be controlled by the entrepreneur or management.

8. Risk exposure

The project is rated excellent if it is considered Less risky, or risks are very minimal while projected profits are more or less assured. Degree of risk can also be assessed in terms of the following factors;

a. The product or service can readily he copied or imitated if the project is found very profitable by others;

b. Competitors who have more resources and expertise may effectively right back if threatened by the project;

c. Changes in customers' and consumers' lifestyle, buying habits, consumption and spending pattern, etc. may take place anytime before the project can service the market;

d. The project may suffer from unforeseen factors such as weather condition, availability of raw materials, technology obsolescence, changes in government policies, priorities or programs;

e. Dependency of the project on imported inputs whether raw materials, technology, skills or other resources.

9. Profitability

At this point in the training, the participants have not yet undertaken business plan preparation. In this case, profitability is a matter of perception or prior knowledge of the participants which may be based on their readings, experience, or observations / feedback from others (competitors, suppliers, consultants, among others). This criterion can be rated in terms of whether the profitability (return on investment) is: a) excellent b) very satisfactory c) satisfactory d) fair e) poor.

10. Cost / Benefit

This tenth factor is practically the summation of all the other nine criteria and gives an overall impression regarding the desirability and feasibility of undertaking the project. The various considerations to be taken into account in rating this factor include;

(a) Whether the benefits of the project in terms of profitability, risk investment requirement, availability of inputs, etc. are worth all the efforts in conceptualizing, organizing, and implementing the project;

(b) The project provides sufficient (tangible and visible) benefits to the community - either through employment generation, backward or forward linkages with other industries or other economic activities; availability of needed products and services, etc,;

(c) The project is economically viable through its own merits and not through artificial government interventions.

11. The Critical Success Factor (CSF)

CSF means a certain factor particular to the identified project which is very important to the success of that specific project. If that certain factor is missing, is inadequate, or is not properly taken into account, it can suggest that the project is not feasible or is likely to collapse.

A project's CSF can be one of the 10 factors above. However, it must be further refined or qualified. For example, not just raw materials, but seasonally of raw materials, perishability of raw materials, lack of standardization of raw materials, unpredictability of availability of supply (perhaps due to import restrictions, infrastructure problem, weather condition, etc.). The CSF concept is also a way of validating the participants' explanation and rating with regard to the 10 micro screening factors.

After the explanation of the whole Micro-screening technique (the meaning of the 10 factors, the rating system, and the CSF), the facilitator can give the participants about 30 minutes to one hour to go through the exercise-asking the participants to micro screen the 10 identified project ideas. The facilitators should be around to provide farther explanations to the 10 criteria or guidance to the participants. To allow more time for discussion or evaluation, the facilitator can prepare in advance the Micro Screen chart according to the number of participants or groups.

After the exercise, the facilitator can ask for class presentation. In a full course setting, volunteer participants can explain the three highest rated projects. In an appreciation workshop or training of trainers, the Micro-screening is a group process; hence each group can be called by the facilitator to take turns in presenting their three highest rated project ideas. In any case, each presentation by a volunteer participant or group should not take more than 10 minutes. The class should be encouraged to ask questions to help crystallize or clarify the participants' ratings. It has been the experience that the other participants always provide constructive feedback: particularly in providing information unknown or unclear to the presenting participant or group. The Micro-screening leads to the SWOT session.

Process of New Product Development

Once a company has carefully segmented the market, chosen its target customer groups, identified their needs and determined its desired market positioning, it is ready to develop and launch appropriate new products. The new product development process for rural market involves four stages, *viz.,* idea generation, concept testing, product development, and test marketing. The product development process remains the same for rural and urban markets except in the case of products specifically used in rural markets.

Idea generation

One aspect that is often ignored in designing product for rural markets is whether the product fits with the rural lifestyle and environment. Therefore, it is important to conceive of a product idea and build a product concept in the rural environment by gaining a first-hand understanding of consumer lifestyle and behaviour patterns through consumer surveys. For example, in rural homes, the detergent bar used for washing clothes is cut into two pieces before use to ensure that it lasts longer due to less melting. But no marketer has addressed this need so far.

Concept testing

Concept testing in rural areas needs to be done in different regions, as needs change from area to area, depending upon the characteristics of particular regions. A concept of low cost dry toilet (requiring little water) promoted by UNICEF was appreciated in water scarce regions of Rajasthan, but was opposed in other regions where water availability is not a problem.

Product development

According to Rogers (2003), Development of an innovation is the process of putting a new idea in a form that is expected to meet the needs of an audience of potential adopters. The packaging of research results is usually done by private companies, so this stage in the technology - development process is called commercialization. Commercialization is the production, manufacturing, packaging, marketing, and distribution of a product that embodies an innovation. Commercialization is the conversion of an idea from research into a product or service or service for sale in the market place. Two or more innovations are often packaged together to facilitate their diffusion because they have a functional interrelatedness, or at least they are so perceived by potential adopters.

Test marketing

The test marketing of any new product is an important factor to decide the success or failure of the product. It becomes critical in the rural context as the chances of failure are often high.

Chapter 8

SWOT ANALYSIS AND MARKET SURVEY

SWOT ANALYSIS

SWOT analysis is a scientific methodology to assess the status of an organization in respect of it probability of sustaining and succeeding. SWOT analysis gives idea about strong and weak points of the organization. The most important advantage of this methodology that it gives forewarning for any likely danger to the organization. Not only that this methodology could be applied to organization, but it could equally be useful for an enterprise or programme.

SWOT is an acronym for Strengths, Weaknesses, Opportunities and Threats in an organization. SWOT analysis is a quality assessment of industrial business and any profit and non-profit organizations management and institutions through a scanning process of the internal and external environment as a part of strategic planning. SWOT analysis is a simple frame work for generating strategies alternatives from a situational analysis. It is applicable to the corporate level and to the business unit level. The SWOT frame work can be designed in short in the following way.

Attributes of organization	Helpful in attaining objectives	Harmful in attaining objectives
Internal attributes	Strengths	Weaknesses
External attributes	Opportunities	Threats

- ***Internal attributes:*** The organization has some inbuilt potentialities which need to be identified and used for the development of the organization. These are the strengths and weaknesses of the organization.
- ***External attributes:*** These are the factors which affect the organization from outside. Cooperation of outside agencies can be obtained for improving the situation of the organization. These are the opportunities and threats of the organization.

1. Strengths

It is positive internal factors that contribute to accomplishing the mission, goals and objectives of an organization. Strength is the basic asset of an organization that would provide competitive advantage for its growth and development. The internal strength of the organization can be assessed by analyzing the environment in which it functions.

Some of the strengths of an organization are;

- Availability of necessary infrastructure,
- Skilled manpower,
- Adequate production capacity,
- Good manufacturing practices, quality assurance and quality control,
- Low cost of manufacture,
- Adequate source of capital,
- Wide distribution network,
- Motivated staff,
- Efficient management,
- Good reputation,
- Good institution building (Rules, regulations, procedures),
- Team building / team work,
- Strong system of monitoring and evaluation.

The qualified and devoted staffs, enthusiastic management team are asset of organization. The relevant programmes undertaken so far might have created a favourable atmosphere for the development of the organization.

2. Weaknesses

It is negative internal factors that inhibit the accomplishment of the mission, goals and objectives of an organization. Weakness is the liability of an organization that can create a state of time and situation specific disadvantage for its growth and development. There may be some short-comings in the organization due to which it is difficult to progress. Some of the weaknesses of an organization are;

- Rising cost of operations,
- Low level of motivation of staff,
- Non-availability of raw materials,
- Scarcity of capital,

- Outdated technology,
- Inadequate infrastructure,
- Shortage of trained technicians,
- Lack of effective coordination,
- Inefficient management,
- Inadequate distribution network,
- Lack of change proneness in organization.

There may be older staff with out-dated ideas, the poor pay scale, long process of appointing new staff, lack of facilities for the staff. In order to develop the organization, it will be necessary to remove these weaknesses.

3. Opportunities

It is positive external factors the organization can employ to accomplish its mission, goals and objectives. Opportunity is the advantage to the organization to grow and achieve its specific objectives in a given situation. The organization has to utilize the opportunities or schemes proposed or in existence outside the organization. Some of the opportunities of an organization are:

- Good location,
- Increase in disposable income,
- Availability of appropriate technology,
- Favourable government policies,
- Availability of different task environment like market information, distribution outlets and media,
- Presence of favourable cultural environment,
- Availability of common basic infrastructural facilities,
- Responsible and responsive surrounding, community or society and particularly the clientele group.

These may be merging with local institutions with reputation, to create excitement of establishing a new institution. The opportunity to enlarge staff expertise to increase range of offering. These are some of the examples for revitalizing the existing organizations.

4. Threats

It is negative external factors that inhibit the organization's ability to accomplish its mission, goals and objectives. Threats are a external situation that blocks the abilities of the organization to grow and develop for reaching its ultimate goal. It is fear for the existence of the organization. If the threats are not properly attended then they will have harmful effect on the organization. Some of the threats of an organization are;

- Shortage of power, water, fuel,
- Rejection by the market,
- Tough competition,
- Fiscal policy resulting into increased taxes, duties, imports reservations, licensing, etc.,
- Obsolete technology,
- Resource crunch,
- Shifting customer tastes and preferences,
- Economic instability with very high ups and downs,
- Prolonged economic depressions,
- Often recurring natural calamities,
- Adversely changing government policies.

Loss of identity, strengths and reputation of organization can be a big threat. There may be risk of losing experienced staff, the ethos of the other organization may become dominant. Such threats need to be identified and attended to avoid the extinction of the organization.

In the management sciences, SWOT analysis is important for understanding the management problems, at all stages, irrespective of the type of organization. The concept is helpful in taking appropriate decision for the development of an organization in a particular operational environment. SWOT analysis has expanded its application in agriculture and allied sectors, apart from trade, commerce, industry, etc.

Examples of SWOT Analysis

SWOT Analysis of Indian Dairy Industry

Sur (2007) identified the strengths, weaknesses, opportunities and threats of Indian Dairy Industry, which shall facilitate entrepreneurship development in this area.

Strengths

- Demand profile – optimistic
- Profit margins - reasonable, even on packed liquid milk
- Flexibility of product mix – good
- Technical manpower - no dearth

- Availability of raw material - abundant.

Weaknesses

- Perishability - need for infrastructure for cooling, pasteurization, sterilization etc.
- Lack of control over yield-use of artificial insemination for upgrading cattle.
- Logistics of procurement - problems due to bad roads, inadequate transportation etc.
- Problems in distribution - need for cold chain.

Opportunities

- Planning Commission has focused on increasing productivity and spoilage management in the farm - consumer chain.
- McKinsey, in the study of processed foods in India, has earmarked processed liquid milk as the highest growth product with good market potential.
- FAO has short listed 26 ethnic Indian milk products for their high potential in the international market.
- Efforts to exploit export potential are already on. Amul is exporting to Bangladesh, Sri Lanka, Nigeria and the Middle East.
- Large number of Indian consumers being vegetarian, vast potential exists for milk products in food habits / cuisine.

Threats

- Environment norms and energy costs warrant a change in packaging material and system.
- Need for immediate implementation of hygiene / food safety norms.
- Milk vendors - this unorganized sector has to be improved.

SWOT Analysis of Vegetable Processing Industry

Maity (2007) identified the strengths, weaknesses, opportunities and threats of vegetable processing industry as follows;

Strengths

- Abundant availability of vegetable based raw materials.

- Priority sector status for agro-processing given by the central government.
- Vast network of manufacturing facilities all over the country.
- Vast domestic market.

Weaknesses

- Inadequate infrastructural facilities.
- Lack of adequate quality control and testing facilities as per international standards.
- Inefficient supply chain due to a large number of intermediaries.
- High requirement of working capital.
- Inadequately developed linkages between R and D labs and industry.
- Seasonality of vegetable based raw materials.

Opportunities

- Large crop and material base offering a vast potential for agro-processing activities.
- Setting up of SEZ (Special Economic Zone) / AEZ (Agri-Export Zone) and Food Parks for providing added incentive ID develop Greenfield projects.
- Rising income levels and changing consumption patterns.
- Favourable demographic profile and changing lifestyles.
- Integration of development in contemporary technologies such as electronics, material science, bio-technology, etc., which offer vast scope for rapid improvement and progress.
- Opening of global markets.

Threats

- Affordability and cultural preferences for fresh food.
- High inventory carrying cost.
- High taxation.
- High packaging cost.

Market Survey

Marketing of product probably is the biggest problem faced by entrepreneurs and the problem is more occur in case of small enterprises. Reduction in production cost and maintaining high quality are the key to successful marketing. While launching an enterprise, first thought needs to be given to marketing. Markets are becoming increasingly dynamic, competitive and risky. To be able to sell the product manufactured, at a fair level of profit, the entrepreneur must possess the ability of making quick decisions in the times of uncertainty. He must know and understand the market where he is going to sell the products eventually. The knowledge of the market gives him the relevant data which is useful in forecasting probable sales levels and in formulating the desired market strategy. Instead of hastily starting a venture, a detailed market survey should be undertaken so as to reduce the risk factor in setting up of a new business venture and also to manage it successfully later on. Projects are undertaken with a view to optimize the sale and maximize profit. Manufactured product must be sold. Production without adequate sale is an invitation to loss of money as well as morale. Absence of sales results in disruption of Production – Sale - Profit chain. It is necessary to know and understand market in which sales are likely to takeplace. Knowledge of market helps in forecasting probable sales and formulation of desired market strategy. Market survey, therefore, plays an important role in successful launching and sustaining an enterprise or production unit.

Meaning of market survey

Market survey is a necessary element of any project. Market survey implies a systematic and rational study of markets to collect information relevant for making decisions concerning marketing. It involves objective and critical analysis of market in terms of its size, location, sales potential, degree of competition, etc.

It may be defined as an objective and systematic collection, recording, analysis and interpretation of existing or potential markets, marketing strategies and tactics and the interaction between markets, marketing methods and current area potential products or services.

The main aim in this exercise is to analyze the market situation *vis-a-vis* your product so as to increase its acceptability and demand. The market survey can be considered while forecasting the future trends. It helps the entrepreneur to get a feel of the market and what goes on in it. A good market survey is an essential pre-requisite of any project. The knowledge of the market enhances the selling capability of an entrepreneur making his project a successful and viable one.

Market survey is a market research tool. Sometimes this is referred to a marketing analysis. It gathers opinion from consumers on the given aspects of product / service under survey and defines the verdict. The trade needs this tool for picking up vital leads to plan new ventures, business expansion, product betterment, amalgamation, merger, diversif-ication etc. It is the survey of the market to find out the supply and demand of a particular product, its equivalent and substitutes, alongwith the price, specifications and other related aspects that are likely to determine or affect its supply or demand, present and prospective. It is very important if you are planning to create a new business, product or service. You need to know who your primary customers will be, what percentage of the market does your competitor dominate, how your competitors find customers etc.

Thus,

- It is systematic and organized collection of all information about products or services.
- Finding conditions prevailing in the market.
- Providing idea about availability and consumption of products and services to the company.
- On the basis of this, company can fix marketing objectives and plan its marketing operations.

Importance of market survey

It is very important to conduct market survey before launching a new project or product line. It is mainly just one way of minimizing risk, so it is similar to buying insurance, which also takes a lot of money. It lowers substantially the chances of investing too much in a project only to realize near the end of it that it is doomed to fail. The second function of market research is to help tailor the product to a target consumer or beneficiary by defining the groups needs and wants and thus improving the level of success for a certain product. Many organizations fail because they ignore market survey and jump into projects without sufficient research. Thus market survey is the systematic and objective identification, collection, analysis and dissemination of information for the purpose of assisting management in decision making related to the identification and solution of problems and opportunities in marketing. The goal of market survey is to identify and assess how changing elements of the market impact customer behaviour. It aims at minimizing the risk in launching an enterprise.

Objectives of market survey

A market survey may have any, or more than one of the following objectives;

1. To mirror the market and its status. It attempts to present a company with reliable, accurate and objective picture of conditions prevailing in the market.
2. To prepare the company for emerging situation by forecasting market trends. Market survey is undertaken by a new company as well as by one which is already established. Market conditions are ever changing, either offering opportunities for

advancements or leading to depressions. Market survey helps in predicting the likely happenings so that the company could ready itself to meet the situations.

3. To fulfill the survey mission. A company may like to obtain reactions to its products or services and undertake the survey work.
4. To establish the saleability and marketability of the product. Market survey finds the demand-supply gap or the likely unfulfilled demand in future. It provides the basis to undertake a project.

Thus, gaining knowledge and understanding of the market is one of the important objectives of a market survey. It may also be undertaken to make an assessment of company performance in the market. Without the periodic appraisal of the distributional arrangements, it may not be possible to know about the reasons for shortages at some places and surpluses at others. Also, it helps in toning the organizational network of the company. The objective of the market survey should be to answer the following questions:

1. What is the size of the market?
2. Who are the customers?
3. What is the demand of the product and how it is being fulfilled?
4. What are the factors limiting the size of the market?
5. What are the pattern of demand?
6. What are the prospects of the demand being increased or decreased in future?
7. How much market share would you like to possess in terms of volume and value?
8. What are the buying habits and motives of buyers?
9. What aspect of the product will appeal to the customer?

   (Unique selling feature and unique selling proposition).
10. Who are the competitors?
11. What are the strengths of their product?
12. What are the past and future trends?

Essential aspects of market survey

Company may propose to have an assessment of

- General market conditions and trends
- Available products, their characteristics and potential

- Demand and supply situation and gap
- Experiences of company products and services
- Performances of company sales strategy and efforts
- Consumer response
- Miscellaneous matters concerning marketing.

Types of market survey

Market surveys can be divided into two types *viz.*, quantitative survey and qualitative surveys.

1. ***Quantitative survey:*** Quantitative surveys mean getting people to answer fixed questions in questionnaires. These questionnaires can be completed over the telephone, in face to face situation, through the post or via internet. Because of the objective is measurement, it is important that all people answer the same question. Changes in wording can dramatically change your measurements. Consequently quantitative interviewers are trained not to stray from the script and good questionnaire design is extremely important.
2. ***Qualitative survey:*** These are about exploring an issue with people. There are no fixed questionnaires and interviewers, use their interviewing skill to draw views and opinions from people using a discussion guide. There are two main approaches to qualitative survey – *depth interviews* (one-on-one interviews) and the *focus group* where the group dynamics means that individuals spark ideas and discussion of each other to explore a topic. It is important that the qualitative interviewer doesn't bias the discussion and lets the interviewees describe things in their own terms. Qualitative work is typically carried out in person, but it can be conducted over the telephone or over the internet also.

Both quantitative and qualitative researches are based on the notion of sampling to identify whom to talk to and the idea of the interviewer as an unbiased observer or collector of information.

Steps in conducting market survey

Market survey should be conducted logically and systematically after deciding about the product to be manufactured. The concrete steps to be followed are as follows:

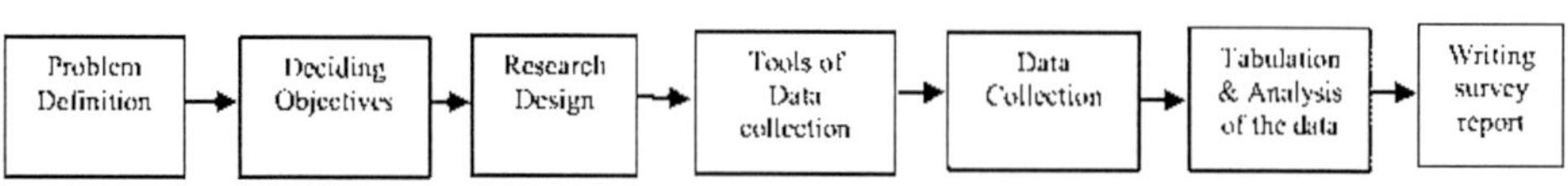

Figure 7: Steps of conducting market survey.

1. ***Problem definition:*** Problem should be such which cannot be solved on the basis of past experience or available information. Unless the problem is critical and well defined, the cost of gathering information may exceed the value of the findings. A market survey is to be conducted only when the problem requires first hand information and the decision will have a critical impact on the future of the enterprise. Problem definition should be precise, and clear. Inaccurate statement of the problem will confuse, leading to wrong results.
2. ***Deciding objectives:*** The second step while doing market survey is to objectively define and spell out the objectives. Based on the objectives and the specific information required, further steps are undertaken. Define the problem accurately.
3. ***Research design:*** The third step after deciding about the objectives and information required, is to work out the details of the study. The sources of information, time and cost involved, resource persons and other relevant aspects will be decided and a detail methodology will be prepared. Survey design is prepared first. It spells out clearly how research objectives are to be achieved. The detailed methodology of the survey is decided including sampling unit and sampling size, type of data needed, methods of data collection, tests and methods of analysis of data, time plan to complete survey, etc.

Data collection is done following two methodologies - 1. Census (population / universe) and 2. Sampling. Census refers to total survey of target market group before a conclusion is reached. It gives a full picture of the object of survey, but is a very time consuming and costly affair. However, a representative group having all the characteristics of the population is selected. This group is called as a sample and a sampling plan is to be prepared with details before collection of information like;

- Sampling unit (who is to be surveyed?)
- Sample size (How many to be surveyed?)
- Sampling procedure (How to select?)
- Contact methods (How to reach?)

Sample is a representative part of the whole that it seeks to represent. It has a fair degree of reliability and accuracy besides being less time consuming than census. The research instrument used is a questionnaire, consisting of a set of question, which can be answered by ticking off one of the given alternatives.

There are two types of data sources, *viz.,* primary and secondary sources.

Primary data sources like manufactures and traders themselves and secondary data sources i.e. the published data should be consulted. Primary source is that source from which the data is collected, from the user's angle, for the first time. It may include the original markets where the product is being sold, or is being sent. Secondary source is one where the data has already been collected and recorded from the primary source. It is important to get the direct experience of market during personal visits and interviews with the persons already established in the business besides collecting the information from secondary data sources.

Primary data are collected from (a) Consumers, (b) Dealers / Distributors / Stockiest / Retailers (c) Manufactures / Producers. While Secondary data are collected from (a) Government departments like SISI, NSIC, SIDBI, (b) Publications / Journals / Literature, (c) Reports and Returns, (d) Case studies and consultants, (e) Research centres.

4. ***Tools of data collection:*** Depending upon objectives and types of survey to be conducted techniques and tools of data collection are decided. Techniques and tools of data collection should be constructed scientifically and systematically. Adequate time needs to be given for preparation of tools of data collection. The tools may be pretested on small sample of respondents for ensuring whether they held collecting appropriate data. Three types of modes of market survey tools are employed, as below. Any one or combination of tools may be used.

   (a) *Interview schedule* – Here the representatives of the survey personally meet the respondents, ask questions, record their replies with the help of the interview schedule. Interview schedule is a set of structured questions. The interviewer himself fills in the schedule while interviewing the people. This method will be helpful when the questions are complex and needs explanation to the respondents. It is versatile and most expensive method.

   (b) *Mailed questionnaire* – Mailed questionnaire is set of questions sent to respondents by mail. The respondents fill in the questionnaire and return it back to the surveyors or company. Mail survey is useful where the respondents are at a distance and spread over a large area and cannot be contacted by personal interviews because of time and cost. Response and accuracy of answers are low. Questions must be simple and clear.

   (c) *Telephone interview* – This has advantages of both the above methods. Questions which are not clear can be clarified; widely scattered sample can be reached quickly at a relatively low cost. Disadvantage is that sample is restricted

to the people who have a telephone. Interview must be short, not very personal.

Sample of Questionnaire for Market Survey of Pickle

i. Do you use pickle? Yes/No

ii. What brand of pickle do you use? ______

iii. Since when have you been using this brand? ______

iv. How much pickles require you per month? ______ gms/month

v. Do you think the price paid by you is

(a) Less (b) Reasonable (c) High (e) Exorbitant

vi. Why do you like this pickle?

(a) The taste is good (b) It is cheap (c) Any other reason ______

vii. Would you want to try another brand? Yes/ No

viii. If yes, will you be prepared to pay a higher price? ______

Types of questions in market survey

i. Dichotomous questions: Division into two parts

Q. Do you like this product? - Yes/No

ii. Multiple choice questions: Possible responses

Q. Why do you like this product?

(a) The taste is good (b) Comparatively cheap

(c) Attractive packaging (d) Trust on the brand

iii. Open end question: Respondent is free to answer

Q. What brand of pickle do you use? ________

5. ***Data collection:*** This step involves actual field work for data collection. It involves planning and supervision. Planning consists of;

- Decisions as to the number of field workers.
- Their respective territories.
- The time period for which the data will be collected

Supervision requires accompanying the surveyors, randomly checking their attitudes to ensure that the data collected are genuine and accurate. Data have to be collected carefully with the help of already developed tools. Checking and cross checking is needed for ensuring the validity of the data collected. The investigators have to probe into the reality as it exists.

6. ***Tabulation and analysis of the data:*** After collection of data, it has to be tabulated and analyzed. The process of tabulation involves the establishment of response categories, the editing of the data and the coding of the data. Editing and reviewing of data make it accurate and unambiguous. Tabulation is the process whereby study data are organized or grouped according to plan so as to facilitate analysis and interpretation. Process of tabulation involves three steps *viz.,* (i) Establishment of response categories, (ii) Editing of the data, (iii) Coding of the data.

Analysis is a systematic interpretation of data so as to arrive at certain conclusions that can be proved by the data. It is the transformation of survey data into inferences and conclusions that can be used for specific purpose

Techniques of analysis

a. Numerical counts.

b. Percentages

c. Measures of central tendency - Mean, median, mode

d. Measures of dispersion – Ranking

e. Correlation and regression techniques

Correlation and regression techniques are widely used for analysis purposes. They may reflect on simple linearity or the multiplicity of the factorial relationship. First, data are organized properly. Raw data are classified, edited and summarized. They are analyzed with various statistical and mathematical tools applied to identify interrelationship and quantitative significance.

7. ***Writing survey report:*** The conclusions drawn from the analysis of data together with major findings and recommendations are presented in the form of a report.

There is no best format of report which will be always best. However in general the format of the report will be as under;

- Summary of conclusions and recommendations.
- Introduction focusing on significance of the survey report.
- Methodology describing the research procedure and sampling plan.
- Sample and its characteristics.
- Detailed findings and observations together with the data arranged in charts and tables.
- Questionnaire and supporting research instruments as appendices.

Chapter - 9

FORMULATION AND FINANCIAL ANALYSIS OF PROJECT

FORMULATION OF PROJECT REPORT

Project formulation involves a number of activities and decisions to prepare blue print of an enterprise to be launched. It involves a series of complex decisions about various aspects of the project like product or services to be provided, availability of raw material, marketing infrastructure, sources of finance and the like. The project report is outcome of project formulation.

A project report provides all the necessary information of the enterprise unit proposed to be set up. It is required by the various developmental agencies / institutions that help setting up of the unit and particularly by financial institutions and banks to provide financial assistance. The report should be prepared in such a way that it is able to provide complete information that may be required by financial institutions for appraising the project. A well prepared project report will not only save time and energy of the banker but also help the entrepreneur, who will not face many objections and queries from the banker.

In India, a lot of published data pertaining to feasibility of various projects is available in a readymade form. The Small Industries Development Organization (SIDO) has brought out good literature covering various aspects of production of many items in the country. The Director General of Technical Development (DGTD) also publishes statistical information for various important items. For example, they publish every year 'Guidelines for industries' indicating those items in which enough scope exists for the creation of further capacity. With so much of ready information available, all that an entrepreneur has to do for starting an industrial unit is;

1. To decide the type and level of industrial activity.
2. To match his financial resources with the required amount.
3. Prepare a project report.

Institutions like Institute for Entrepreneurship Development (IED), Small Industries Service Institutes (SISI), State Industrial Development Corporation (SIDC) or State Small Industries Development Corporation (SSIDC) help in preparation of project report and later on recommend them to the banks. Besides these institutes, several commercial banks also render assistance in this respect.

Content of Project Report

We will now briefly discuss the various factors which should be presented in a Project Report. However, these guidelines are meant for small scale industrial projects only and hence will not include any sophisticated technique being used for preparing project reports of large scale enterprises. Within the small scale industries too, all the information may not be relevant for all units. The size and type of the unit will be the deciding factors for the amount and type of the information to be furnished in the project report.

The following are broad heads under which complete information on relevant aspects should be included in the project report:

1. General information
2. Project description
3. Market potential
4. Capital costs and sources of finance
5. Assessment of working capital requirements
6. Other financial aspects
7. Economic and social variables
8. Other details

1. General Information

To begin with, some information of general nature should be provided. The following are some aspects;

a. ***Bio-data of promoters*** - It includes name and address, the qualifications, experience and other capabilities of entrepreneur. If there are partners, state these characteristics of all participants individually.

b. ***Industry profile*** - A little reference of analysis of industry to which the project belongs, e.g., past performance, present status, way it is organized, the problems, if any, etc.

c. ***Constitution and organization*** – The information has to be given about following points. The constitution and the organizational structure of the enterprise. In case of a partnership firm, whether registered with Registrar of Firms. Whether a registration certificate from the Directorate of Industries / District Industries Centre has been obtained or will be supplied later on.

d. ***Product details*** - The utility of the product and range of the product to be manufactured. One could even provide the product designs / drawings alongwith and a mention of the advantages of proposed products over its substitutes.

2. Project Description

A brief description of the project covering the following aspects should be given in the project report.

a. Site

Location i.e. town, locality, street number etc. Whether owned or leased hold? Whether site is in approved industrial area? Is it suitable to the type of enterprise being planned? The open / covered area available / needed should be mentioned. If location is in residential area then the copy of No Objection Certificate from Municipal Authority should be attached.

b. Physical infrastructure

Availability of physical infrastructure consisting of the following items;

Raw material – Whether imported raw material is required? If so, whether the license has been obtained? Which are the source of raw material and what is the probability of getting it on a continuous basis at fair prices?

Skilled labour – Whether skilled labour is available in the area? If not, what arrangements have been made to train the labour in various skills?

Utilities - Availability of required utilities *viz.,* power, fuel, water, etc. should be provided.

Power – Inadequate supply of electricity or its high unit cost in an area may become major constraints in running a project. The project report should contain the information regarding the power requirement, the load sanctioned, stability of supply of power and the price at different consumption levels.

Fuel - Whether other fuel items like coal, coke, gas or oil are required and if yes, then state their availability position

Water – Water is an important factor for projects like brewery, tannery, ice plant, and soft drinks, chemicals. The source and the quality of water in such cases should be clearly stated.

c. ***Pollution control:*** Most industrial plants produce waste material or emissions that may create significant problems to the environment. The emission may be of various types such as (a) gaseous (smoke, fumes etc.) (b) physical (noise, heat vibration etc) or (c) liquid or solid discharge through pumps and sewers. State clearly the aspects like scope of dumps, sewage system and sewage treatment plant.

d. ***Communication system:*** Availability of communication facilities such as telephone, telex, internet, e-mail, etc should be stated in the report.

e. ***Transport facilities:*** The distance over which the material inputs will have to be transported and the available as well as potential means of transportation should be stated together with expected bottlenecks, if any.

f. ***Other common facilities:*** Availability of facilities such as electrical repair shops, machine shops, welding shops, etc.

g. ***Manufacturing process:*** The details of production and process involved should be clearly stated. Also state the period of conversion from raw material into finished goods. A process flow chart should be presented.

h. ***List of machinery and equipment:*** A complete list of items of machinery and other equipment indicating their type, size and cost should be stated. Sources of supply of capital, equipment and the construction services should also be given. Check machinery / equipment for each of the above stated processes are ensured. If not, explain how such processes will be got done.

i. ***Capacity of the plant:*** The installed licensed capacity should be stated. Also state whether unit will run on single, double or triple shift basis.

j. ***Technology selected:*** Is it up-to-date and appropriate? Which other units are using same technology? What are the results of this technology? How is the required know-how proposed to be arranged?

k. ***Balancing of plant:*** While stating the stages of production, also state whether capacity of various plants at different stages of production is sufficient? Balancing equipment required at later stage and the consequent increase in capacity should be assessed.

l. ***Quality control / testing and inspection:*** Whether some system has been designed to check quality on continuous basis? Obtaining quality marks like ISI, Q or AGMARK help in creating confidence among consumers.

m. ***Research and development:*** Besides the quality control, whether any cell to study improvement of quality is proposed to be formed in the enterprise?

3. Market Potential

The following aspects relating to market potential should normally be covered in the project report.

a. ***Demand and supply position:*** State the data regarding total expected demand of the product and present supply position. How much of this gap will be filled up by the proposed unit?

b. ***Price expected to be realized:*** An estimate of the price expected should be furnished to assess the margin of profit. A comparative statement of competitors selling price would be helpful.

c. ***Marketing strategy:*** What strategy for selling the products is proposed to be followed? Whether any arrangements have been made with reputed suppliers and distributors for lifting production? Sometimes, particularly in electrical goods, owners of reputed brands may enter into contract to lift the entire production and sell it later after putting their trade mark on it.

d. ***After sales service:*** In some items it is very vital. Even due to a loose screw or snapping of a wire, the customer may find the instrument either not working or working improperly and without after-sales service, the product will get a bad name. Normally it has been found that money spent on after-sales service by a manufacturer is repaid many times over in the long run.

e. ***Seasonality factor:*** Whether the items have seasonal fluctuations in sales? If so, the arrangements made for warehousing or stocking of the goods in off-season should be stated.

f. ***Transportation:*** Whether the unit will depend for the transportation of goods by public carrier or by own transport? In case of own transport, mention cost and assistance required.

4. Capital Costs and Sources of Finance

An estimate of the various components of capital items required by the unit should be given in the report. These components may be:

- Land and building
- Plant and machinery
- Installation costs

- Other miscellaneous assets like furniture / fixtures, vehicle, tools etc.
- Preliminary and preoperative expenses
- Contingency cushion against price rise / unforeseen expenses
- Margin for working capital

Besides the cost factors, the report should include present probable sources of finance. The sources of funds should equal the cost of a project otherwise the project cannot be set up in full. The sources would include owners' fund together with loans and deposits raised as well as the loans expected from financial institutions/banks. The estimation of funds for the cost factors involved should be realistic and correct. Many units run into serious financial problems because of inadequate estimate of funds requirement.

5. Assessment of Working Capital

Planning for working capital requirement is equally crucial for an entrepreneur. While estimating the capital costs, margin for working capital was taken into account. Any unit will be able to function only when adequate working capital funds are available. In other words, at the initial stage itself, the estimate of working capital requirement should be made and shown alongwith the total cost of the project. Sometime back formats for working capital assessment had been designed. As such if the entrepreneurs present their estimates in those prescribed formats, it will save time and energy for them as well as for the banker.

It has been generally noticed that the entrepreneurs present the working capital requirements in their own way which is ultimately recasted by the banker. This results into wastes time and creates problems. Hence, if they project their requirement in the prescribed way, it will minimize objections from the banker's side.

6. Other Financial Aspects

One of the objectives of setting up a project is to earn profit or a livelihood. Besides the project set up must be able to retrieve the investments made within its life cycle. This would be possible only if products taken up for production are adequately profitable. This would require preparation of a projected profit and loss account which would indicate likely sales revenue, cost of production, allied cost and profit. These estimates, especially the likely sales revenue, should be made on a realistic basis. A projected balance sheet and cash flow statement would also have to be prepared to indicate financial position and requirements at various stages of the project. After all the smooth functioning of the unit necessitates availability of adequate funds for various commitments.

Next the Break-Even analysis must be presented. Break-Even point is the level of production / sales where the industrial enterprise shall make no profit no loss, it will just break even. This facilitates knowing the gestation period and the likely moratorium required for repayment of loans.

$$\text{Break - Even Point (BEP)} = \frac{F}{S-V} \times 100$$

Where, F = Fixed cost, S = Sales projected, V = Variable costs

The break-even point thus will show at what percentage of projected sales the unit will break even. It is also a good idea to calculate and indicate the following ratio;

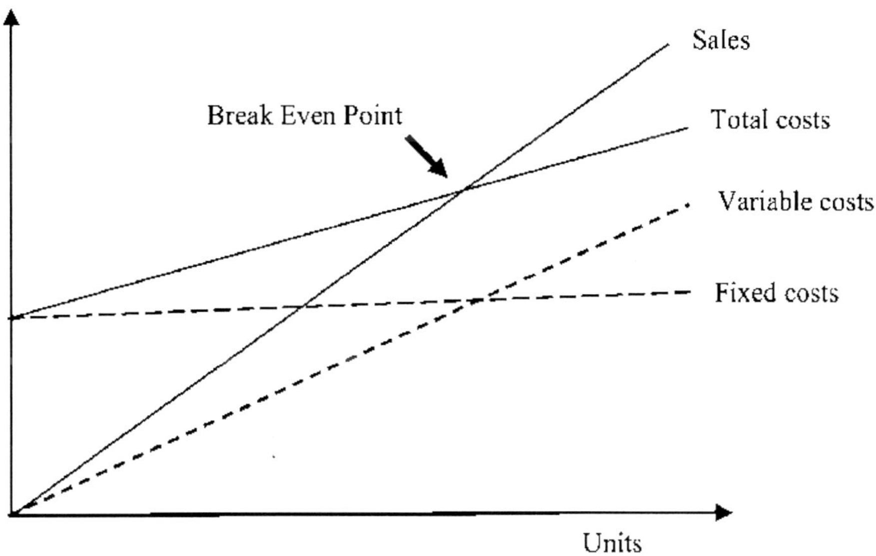

Figure 8 : Break–even point chart.

$$\text{Profitability ratio} = \frac{\text{Net profits}}{\text{Sales}} \times 100$$

$$\text{Return on investment} = \frac{\text{Net profits}}{\text{Capital employed}} \times 100$$

$$\text{Debt : Equity ratio} = \frac{\text{Debt}}{\text{Equity}} \times 100$$

$$\text{Debt service} = \frac{\text{Net profit after tax} + \text{Depreciation} + \text{Interest (for one year)}}{\text{Installments} + \text{Interest (For one year)}}$$

7. Economic and Social Variables

What will be the abetment costs, i.e., the cost for controlling environmental damage (e.g., pollution)? The abatement cost will constitute the value of the additional engineering and technology needed for treating effluents and emissions. Whether project will have socio-economic benefits, which of following are a few examples;

Employment generation – The number of persons proposed to be employed *vis-à-vis* employment situation of that area may be mentioned.

Import substitution – The manner in which it is planned to be achieved and the amount of benefit expected may be mentioned.

Ancillarasation – Whether the unit will need sub-contracting functions of such type that ancillary industrial units may be promoted to meet them?

Exports – Quite likely products proposed for manufacture may be exported in full or in part.

Local resource utilization – Certain local resources which are presently a waste may be usefully utilized upon the project going on stream.

Development of the area – How the establishment of the unit will bring on overall development in the area of its operation?

8. Other Details

a. ***Project implementation schedule*** – Preferably a PERT / CPM chart can be appended to the project report. If this is not feasible then in a tabular form likely dates of completion of the following activities can be simply mentioned;

- Acquisition of land
- Registration of unit
- Bank loans
- Construction of building
- Power connection
- Ordering plant/machinery
- Installation of plant / machinery
- Recruitment of workers
- Training of workers
- Ordering of raw material
- Procurement of raw materials
- Trail run

- Commercial production

b. ***Plant layout*** - If possible, a copy of the plant layout can also be furnished in the project report. This will assist determining sufficiency of area for present and future expansion requirements.

c. ***Annexure*** - The following is an illustrative list of appendices of a project report, which may be attached alongwith a project report.

- Equipment required
- Details of preliminary expenses
- Estimates - Cost of production and profitability
- Labour and staff requirements
- Break-even point
- Working capital requirements
- Debt-service coverage ratio
- Cash flow statement

FINANCIAL ANALYSIS OF PROJECT

Financial analysis is necessary for the entrepreneur putting up the project, the financial institutions and bankers providing loans for the project, shareholders and other organizations participating in the equity of the project, and all those involved in appraisal of the project. The financial analysis is a prerequisite to assessing the viability of the project and there are many cases when a project has become sick due to incorrect financial appraisal at the project stage, shortage of working capital and inherent weakness of the project from financial considerations.

The financial analysis of projects is designed to know that whether estimates are properly done for the cost, capital, sources of finance, break-even point, loan repayment period, cost of production and profitability. Various financial ratios like profit to sales ratio, debt equity ratio, profit to total capitalization ratio, etc.

[A] Cost of Project and Means of Finance

In order to decide the financing pattern and economic viability of the project, it is necessary to estimate the cost of the project accurately. Upon this depends the projections of cost of production, cash flow and profitability. If the cost of project is underestimated and there is over-run, we will have to raise additional resources which may delay the project. It may also affect project viability. If the project cost is over estimated, then one is making higher commitments and incurs unnecessary commitment charges and interest which is not in the interest of the project.

Cost of Project

Following are the major components of the cost are;

1. Land and site development
2. Cost of laying roads, fencing and gates.
3. Various buildings like main factory, administrative building, stores, utilities, canteen, essential housing, etc.
4. Plant and machinery like cost of equipment, spares, duties and taxes, packing and forwarding, freight, foundation, erection and commissioning.
5. Technical knowhow like fees, expenses on drawings, consultancy fees etc.
6. Training and deputation of foreign technicians.
7. Cost of miscellaneous fixed assets like furniture, office machinery and equipment, vehicles, laboratory equipment, telephone, internet, workshop equipment, etc:
8. Preliminary expenses related to formation of company, cost of raising capital by public issue.
9. Preoperative expenses means expenses during preoperative / construction period, i.e. period before start of commercial production, *viz.,* establishment expenses like salaries, rent, travelling and conveyance etc, expenses on interest on term loan, expenses on start up and trial run.
10. Contingencies because of unforeseen factors, fluctuation in foreign currency value and inflation. Usually 5% to 15% contingency provision is made.
11. Margin money for working capital can be estimated with the inventory quantum for materials, consumables, stores, work in process and finished goods.

The bank finance for working capital would be available with a margin of 25 to 40% depending on the scale of the project and the risk value of the security *viz.,* inventories or bills.

The margin money for working capital = Total working capital - Bank finance.

This has to be financed by long term resources and is therefore included in the cost of the project.

Means of Finance

After the total cost of project has been estimated, one has to decide on the pattern of financing the project. To meet the cost of project, the following means of finance can be considered;

Generally sources of business capital can be divided into two parts.

1. Owned funds – Share capital, and ploughing back of profit
2. Borrowed funds – Debentures, term loans, public deposits, commercial banks, and intercompany deposits

(1) Sources of own funds: The capital invested by owners themselves in the business unit is called own capital. Joint Stock Company can collect share capital as owned capital as well as profit can be reinvested in a business unit. Broadly speaking there are two sources of owned capital such as share capital and ploughing back of profit.

a. Share capital - Company act, 1956 provides that a joint stock company can issue shares to collect owned capital. A silent feature of joint stock company is that it can collect large capital through share capital. Share capital of a company is divided into very small parts and each small part is called share. A person holding a share is called share holder.

The company act, 1956 provides that share capital of a company, limited by shares, shall be of two kinds; (i) Equity share capital and (ii) Preference share capital.

(i) Equity share capital – It represents the company's assets and a share of any profit earned on those assets after others claims have been met and the equity share holders are the owners of the company – they purchase the equity shares, the money of equity shares is used by the company to buy assets and the assets are used to earn profits, which belongs to equity share holders. This is permanent capital brought by the owners / promoters or project with no liability for repayment or any obligation for paying dividends. Hence the company needs not to return equity capital, except during its winding up. The equity shareholders bear the risk of ownership and also enjoy the rewards of profits.

(ii) Preference capital – It is a hybrid type of security, as it has features of both ordinary shares and bonds. This share capital is called preference share capital because preference share holders have preferential rights of assets and dividends. In preference capital, a fixed dividend is paid (this way it is similar to loans/debt) and the dividend is taxable (this way it is similar to equity). Preference share holders get their dividend before equity share holders can receive their dividends and in the event of winding up, they have a claim on available assets before ordinary share holders. The dividend on preference share is fixed.

b. **Ploughing back of profit:** If a joint stock company earns ample profit, it invests some portion of such profit a reserve nature, instead of distributing all profits as dividend. Such profit invested as reserve is used in the business operations of the company, which is termed as 'ploughing back of profit'. Such profit can be used either for expansion or modernization of the company. Legal provisions and limits about ploughing back of profit have been made by govt. of India. Due to higher ploughing back of profit, a company can survive even during recession time and can face easily financial crisis. So it increases company's financial soundness.

(2) **Sources of borrowed funds / capital:** In the present era, each and every business activity needs huge capital investment. Such large capital requirements cannot be fulfilled by only owner's capital. Hence entrepreneur uses borrowed funds also. Important sources of borrowed funds are; (a) Debentures, (b) Term loans, (c) Public deposits, (d) Commercial banks, (e) Inter company deposit.

a. **Debentures** - These are long term debt instruments which are protected by a charge on the properties of the issuing company. If a company needs long term capital but the company is in a position to redeem such capital after a fixed period, it can issue debentures. Debenture is an acknowledgement of debt, given under the seal of the company and containing a contract for the repayment of principal sum at a specified date and for the payment of interest at fixed rate per cent until the principal sum is repaid and it may or may not give the charge on the assets to the company as security of the loan. A debenture is a kind of document acknowledging the money borrowed, containing the terms and conditions of the loan, payment of interest, redemption of the loan and if any security offered by the company.

b. **Term Loans** – It is also referred to as term financing. These are long term loans carrying fixed rate of interest and repayable in 5 to 10 years in equal semi-annual installments. Term loans are provided by banks, State Financial Institutions. State Development Corporations, All India

Financial Institutions like IDBI, ICICI, IFCI, UTT, LIC, GIC. They provide term loans to business ventures for financing their medium term and long term assets as well as permanent working capital. The interest charged by the lending institution depend upon the general level of interest rate, financial position of the company, risk perception of the lender and manner and period of repayment. The term loans are sanctioned on a thorough appraisal of the investment proposal.

c. **Public deposits** – Public deposits are very important sources of short term and working capital. According to the companies rules, 1978 a company can accept public deposits. Public invest their savings as deposits in the company. The term of public deposits is from 6 months to 36 months and the rate of interest is generally 10 to 14 %. Due to higher rate of interest on public deposits, presently large portion of our society deposit their money with well known large scale industries. To get higher rate of interest, public deposits are more elastic. Because of company can return such deposits on maturity period and can become free from the burden of interest. The limit of accepting public deposits is up to 25% of paid up capital and free reserves.

d. **Commercial banks** – In India, commercial banks provide assistance for working capital requirements and other short term needs of an enterprise. Commercial banks collect deposits from public generally for short term and such savings are provided as financial assistance to entrepreneurs. In India commercial banks provide assistance in various forms like; loan, over draft, cash credit, bill discounting, export credit, guarantee etc. Commercial banks provide assistance as per the guidelines of RBI. Presently RBI has been encouraging the commercial banks to provide term finance to small ventures, particularly small scale industries.

e. **Inter company deposit** – In India, there are many groups of management who are managing various branches like Tata, Birla, Reliance, Bajaj group etc. Such group has ownership of many companies. So in such a group of companies, if there is surplus capital fund in a company, this is lended to the company having its need for short time. Sometimes a company may deposit capital fund in other company of the same group in the form of 'call deposit'. So by inter-company deposit, a company can get financial assistance with lower interest rate and without legal formalities though Government of India has placed many restrictions on inter-company deposits.

(3) **Deferred Credits** - Many times the suppliers of machinery provide the facility of deferred credit under which payments for purchase of machinery can be made over a period of time.

a. ***Capital subsidy and development loans / sales tax loans*** - The government provides capital subsidy to industries set up in notified backward / hilly areas. Some state governments also provide development loans/sales tax loans for development of an area.

b. ***Unsecured loans and deposits*** - The promoters sometimes have to provide for unsecured loans to bridge the gap between the promoter's contribution required and the capital, which can be raised by the promoter. These loans are subsidiary to the institutional loans.

c. ***Promoters contribution*** - The promoter's must have a stake in the project. For this purpose, a certain percentage of the cost of the project should be financed by the promoters. The promoter's contribution is decided on the basis of the project, location, background of promoters. The margin is low for technical and women entrepreneurs.

[B] Cost of Production

In arriving at the cost of production of product, we have to look at the various factors as outlined below;

1. ***Material and consumable stores*** - (i) Check the requirement of material per unit of production. Add for yield, wastages, and rejection. This information can be taken from collaborators, industry experience, and consultants. Based on above, calculate the total requirement for the year. (ii) Compute the price of materials after adding freight, insurance, taxes and other expenses.

2. ***Utilities (Power, fuel, gases, water, etc)*** - This is based on technical data for equipment and manufacturing process. The cost is determined as per tariff structure of Electricity Board (for power), and prevailing prices of fuel (refer Indian Oil Corporation), gases etc.

3. ***Repairs and maintenance*** - The expenses on this head will increase over the years as machinery gets old. One has to consider repairs and maintenance expenses on building, plant and machinery and miscellaneous assets.

4. ***Rent, insurance, taxes*** - Rentals depend on the area, type of space and location. The cost of insurance depends on nature of property to be insured, the risks to be covered and the value of property. The financial institutions and banks insist on insurance of assets charged to them as security.
5. ***Wages and salaries*** - (i) Check the requirement of workers, supervisors, staff; engineers and managers. (ii) Check the wage structure in the area of the plant and office. (iii) Provide for annual increases in wages and salaries towards inflation and increments.
6. ***Administrative expenses*** - Directors remuneration, salary of staff and general management, audit fees, travelling and conveyance, postage, telephone, etc., are included here.
7. ***Sales expenses*** - Expenses on marketing, distribution, sales promotion, discounts, commissions, advertisement, salaries to sales persons, packing and forwarding, etc.
8. ***Royalty*** - This is an expense to be paid to the technical collaborator on a fixed cost per unit basis or as a percentage of selling price.
9. ***Excise duty*** - This is leviable on the product by central government and is charged on the ex factory price.
10. ***Financial expenses*** - (a) Interest on term loan. (b) Interest on bank borrowings for working capital.
11. ***Depreciation*** - Depreciation can be calculated on straight line or written down value basis, for all assets except land. The rates of depreciation are different for different assets depending on their functional life.

The sales revenue is calculated based on product mix, capacity utilization, production level, adjustment for opening and closing stock, and the selling price net of the duties and taxes.

After deducting the expenses from the sales realization, the profit before taxation is worked out. Tax is calculated according to the tax laws and tax concessions available are considered while computing tax level. The profit minus tax is the net profit after tax. The company can declare dividends and make repayments of term loan from profit after tax. The balance goes towards reserves of the company.

[C] Financial Projections

Cash flow statement: Cash flow estimates are prepared to ensure that the unit will have necessary cash with it and it will not face liquidity problem. The cash flow estimates include sources of funds and their dispositions (uses). While profitability estimates are prepared only from the year in which the unit is likely to commence production cash flow estimates are necessary for the construction period also to ensure availability of cash according to the requirement of the project. There are a number of sources of funds which have been described under means of finance earlier in this chapter.

The uses of funds include the following:

a. Capital Expenditure
b. Preliminary Expenses
c. Current Assets - increase in level of inventories & book debts.
d. Decrease in term loans/debentures/deferred credit
e. Decrease in unsecured loans/deposits
f. Repayment of bank borrowings
g. Interest
h. Dividends
i. Taxation

The difference between the sources and uses of funds indicates the net surplus or deficit. Accordingly, one has to take a business decision regarding management of funds.

Projected balance sheet: Projected balance sheet is prepared on the basis of profitability statement and cash flow statement. The position of share capital, term loans, sundry creditors, bank borrowings, current assets etc. is determined at the end of each year according to the movements shown in cash flow and profitability estimates. Fixed assets are taken after deducting depreciation provided in profitability estimates. Preliminary expenses are taken after deducting the amount which is already written off from the expected profits of the unit. Cumulative surplus shown in profitability estimates represents the position of reserves at the end of each year, closing balance shown in cash flow estimates represents the position of cash and bank balances at the end of each year. The balance sheet is a snap shot picture of the financial position at the end of each year and the total of its assets side will be equal to the total of its liabilities side.

[D] Financial Analysis

Break even analysis: Break-even point is the level of production at which the unit neither earns a profit nor incurs loss. Break Even Analysis aims at finding the behaviour of costs with volume of production, the production/sales level at which the firm breaks even, the sensitivity of profit to variations in cost, price and volumes, etc.

There are two types of cost of production - fixed cost and variable cost. Fixed cost is incurred irrespective of the level of production. For example, depreciation, interest on loans, some portion of wages and salaries, power, rent, administrative overheads, insurance, selling overheads can be considered as fixed cost.

Variables cost or marginal cost varies with production. For example, raw materials, working capital interest, selling commission, some portion of utilities, wages and salaries, packing charges, excise duty can be considered as variable cost.

$$\text{Contribution} = \text{Sales Revenue} - \text{Variable Cost}$$

The contribution helps unit to recover its fixed cost. The level of production at which the contribution recovers entire fixed cost is called break-even point.

Break-even point indicates the level of production which is necessary to avoid losses. Sales above break-even point will bring profits. Break-even point should be low for units likely to have difficulties regarding marketing or non availability of materials or shortage of power or any difficulty in achieving utilization of full capacity.

An entrepreneur will prefer to close his plant when variable cost is more than sale price and if the situation is not likely to improve in near future. By closing plant, only the fixed cost will be lost but running the plant one will lose the fixed cost as well as the excess of variable cost over sale price.

Payback period: The payback period is the time required to recover the initial capital outlay on the project. The shorter the payback period, the more desirable the project.

Profit to total capitalization ratio: This is worked out by dividing profit (before taxes with long term debt interest added back) by total capitalization (i.e. equity plus long term debt). This measures the earning power of both owned and borrowed capital.

Debt service coverage ratio (DSCR): Financial institutions which provide the bulk of long term finance for industrial projects evaluate the financial viability primarily from the measures of DSCR and Internal Rate of Return.

$$\text{DSCR} = \sum_{i=1}^{n} = \frac{\text{Profit after tax} + \text{Depreciation} + \text{Interest on Term Loan}}{\text{Term Loan Installments} + \text{Interest on Term Loan}}$$

Where 'n' is the number of years over which the loan has to be repaid.

A DSCR of 2 is considered satisfactory.

Net present value (NPV): The net present value is the sum of the present value of all the cash flows associated with the project. One may accept the project if NPV is positive and reject if NPV is negative.

Internal rate of return (IRR): The internal rate of return of a project is the discount rate which makes net present value equal to zero. In the net present value method (NPV), we assume that the discount rate (cost of capital) is known and determine the net present value of the project. In IRR calculation, we set the NPV to zero and determine the discount rate (IRR) which satisfies this condition. The calculation of IRR is one of trial and error. We try different values of discount rate till we arrive at one which makes the present value of cash inflows (profit after tax, depreciation, salvage value) equal to the present value of cash outflows (capital) expenditure, increase in current assets, calculated over the life of the project.

Debt-equity ratio: The financial structure of the company should reveal a satisfactory balance of owned funds (equity) and borrowed funds (debt). Prudence demands that the ratio should be between 1 to 2, certain institutions have debt equity norms for different categories of borrowers and may be as high as 3:1 for tiny sector, to a low of 1:1 for large industries.

Current ratio: This ratio gives the relationship between current assets and current liabilities. Its purpose is to indicate whether the project can pay off the current liabilities as they mature and whether it can withstand reverses by the strength of its liquid position,

Ratio of sales to capital employed: This ratio is a broad indicator of the efficiency and economy observed in the use of capital.

Sale to receivables ratio: This ratio discloses the collection problems of credit sales. A ratio of 1:2:1 implies book debts equivalent of sales of one month. A fall in this ratio is considered unfavourable as it may amount to a larger collection period and reflects on the quality of book debts. Thus if the ratio drops to 3:1, it shows 4 months uncollected debts and if the business custom is to extend only 1 months credit, the matter of higher collection period has to be investigated.

Profit to sales ratio: As the profits accrue only from sales, this ratio provides a yard stick of the efficiency of production and a measure of margin on sale price.

Chapter - 10

MARKETING AND ADVERTISING PRODUCTS

Marketing means different things to different people. To some it is a selling activity, others see it as advertising. While still others may give a different interpretation. Each of these is a part of marketing, but not the total package.

Marketing is not an activity which should be seen in isolation. It really means taking a view of the whole enterprise and its ultimate objectives. One of your first tasks as an entrepreneur is to make correct decisions about the kind of products and services that can he produced and sold to customers. Marketing will help entrepreneur in taking the right decisions.

Concept of Marketing

Marketing consists of series of activities involving the flow of goods and services from the producer to customer. The objective is to satisfy the needs and wants of customers and enables your enterprise to make a profit. Marketing activities include:

1. Finding out what people need and want.
2. Developing the product or service to meet this need.
3. Pricing the product.
4. Making the product available at places where people can buy them.
5. Promoting the product to inform people about it and retain the interest of buyers.

Each stage needs to be carried out effectively. It is important to remember that customers usually have a choice of whether to buy or not to buy a given product. In making a product choice of people often consider the benefit that the product has to offer and how well it meets their needs. This focus on the buyer or consumer is called Marketing Orientation. This will help in two ways:

1. In providing a product that the customer wants.
2. Make business grow bigger.

Market Assessment

To operate enterprise successfully, you need to collect information that will enable you to make the right marketing decisions. The information required will be;

- What product to produce or sell?
- Who the buyers are and where they are located?
- When do they buy and how much?
- Why do they buy?
- How to attract buyers?
- What price, customers are willing to pay?
- Who are the competitors, and where they are located?
- How do they operate? What are their strengths?
- What substitutes are available for the product?

To collect this information talk to;

- Sellers of similar product.
- Other producers.
- Buyers or people in the community.

You can also visit a shop or market where similar products are being sold and watch how people are buying. Based on this information and assessment of your own skill and resources you should be able to decide: the product to sell, where and how to attract buyers and how to compete effectively with others.

As a part of your research, you should test a small quantity of the product in market. This test marketing will help you in finding out if the product is accepted or further improvement are required, before you go into full scale production.

Elements of the Marketing Mix

According to Chhabra (1998), all marketing activities can be divided into four elements known as the Marketing Mix - Product, Price, Place, Promotion, They are often called the 4P's of marketing on account of the first letter of each of these words. On the basis of information collected earlier, you are now ready to take a decision on each of the 4P's :

1. **Product:** Product is anything that is offered to someone to satisfy a need or a want. It may be a physical object or a service.
   - Develop a product in response to what the customers want.

- Decide the shape, colour, package, brand name, quantity and quality the buyers want.
- Decide whether you can make your product like more than other sellers.
- Determine the availability of materials for making the product.

2. **Price:** Price is the monetary value of your product. You should be able to sell at a price that will attract buyers. To arrive at the right price, you should consider;
   - The cost of production of product.
   - How people buy and how much are they willing to pay for your product?
   - The price of other sellers (competitors).
   - How different is your product (uniqueness).
   - The profit margin you want.
   - Whether your product is seasonal?
   - Use of special prices (discounts) for some customers.

You may be able to sell at a slightly higher price if your product is different from that of other sellers. If a customer buys in bulk you can give him / her a quantity discount.

3. **Place (Distribution):** This means how and where you make the product available to the customers. Here you should consider :
   - Delivery of the product - Direct or through middlemen.
   - Transportation.
   - Place at which to sell.
   - Storage facilities and cost.

There are many ways in which you can reach customers as shown in the figure below:

Stated in words it means;

- You can sell directly to consumers yourself or through your sales people.
- You can sell to other retailers.
- You can sell through wholesalers, who resell to retailers from whom consumers finally buy the product.

The decision to use middlemen should be carefully made because it involves some cost. The longer the channel, the higher the cost which means a higher price to the customer. This may affect demand for your product. This does not mean that middlemen should not be used.

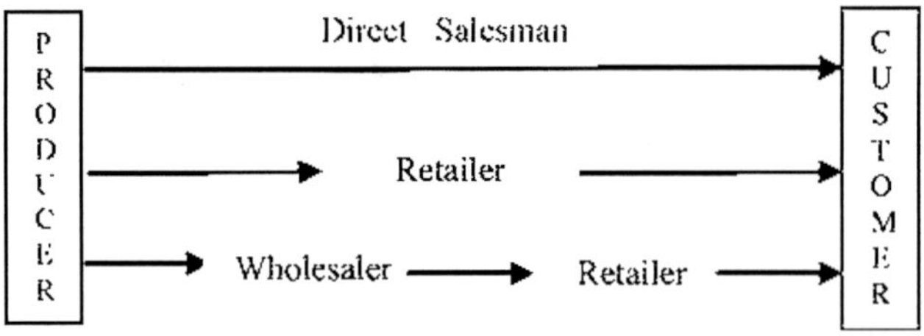

Figure 9: Ways of delivery of the product.

Selection of channel depends upon

- Nature of the product e.g. perishable like milk, fruits, vegetables, meat, fish, expensive items like jewellery etc. In such cases, shorter and direct channels should be used.
- Reliability of the channel and its image.
- Cost effectiveness.
- Speed of reaching customers.
- Location of the target customers.
- Storage facilities.

Identify distribution points. These could be -

- The local market
- Departmental store.
- Home
- Door to Door.
- Corner shop.
- Fairs / Melas.
- Cooperatives.

The choice of these points will depend upon

- Nearness to buyers.
- Sufficient demand at the location.

4. **Promotion:** This is the last element of marketing mix. Promotion means marketing communication. It involves any activity which influences people to buy your product. These may be in the form as under;
    - Advertisement through mass media such as newspaper, posters, radio, internet, etc.
    - Display of the product outside the shop.

- Arrangement of the products inside the shop.
- Giving products on discount or free samples as gifts.
- Use of jingles, bells, music to attract attention.
- Use of signboards
- Door to Door visits / sales.
- Distinctive brand naming the product

While creating marketing messages you must point out the benefit, the customer will obtain from the product, and not its features. Instead of saying 'Our apples are big and fresh', we should say 'Our apples are delicious'.

Each element of the marketing mix is important, but the emphasis given to each one will vary with the type of the product. Each product must have an appropriate mix of 4P's.

ADVERTISING PRODUCTS

Major success of an enterprise depends on the marketing of the products. Product must be liked and preferred by customers, for which they should have some knowledge of the branded product. Advertisement plays a role of informing the potential consumer about the product. It also makes the attitude of customers favourable, persuades them and makes environment congenial for sale of the branded products. It catches market for the products.

Meaning and definition of advertisements

The word 'advertising' is derived from the Latin word *'advertese'* which means *'to turn (the mind) to'*. It is turning one's attention to the message of either a product or a service.

According to Concise Oxford Dictionary, the verb *'to advertise'* means 'to make generally or publicly known; describe publicly with a view to involving sales.' The Webster Collegiate Dictionary gives the meaning of advertising as 'the action of calling something to attention of public by paid announcement'. The Encyclopedia Britannica defines advertisement as 'a form of paid announcement intended to promote the sale of a commodity or service, to advance an idea or to bring about some other effect, desired by the advertiser'.

Broadly speaking, advertising does turn the attention of the public to a commodity or service and in the broad sense it might be said that anything that turns attention to an article or service might be advertising. In this way, advertising began when someone had some product to sell and at the same time there was someone who wanted to buy that particular product. It becomes a process of 'selling' and 'buying' through public announcements. Advertising touches the lives of almost every person, almost every day. It provides information about products, services and ideas and influences the choices one makes. It often guides us in our purchase and actuate our thinking on matters of personal, social and economic importance. Today, advertising has become a part of the economic and social fabric of the nation. However, the power of advertising has been employed in both good and bad ways. Advertisement differs from publicity because it is paid for directly and its sponsorship is almost always clearly identified.

The American Marketing Association defined advertisement as 'any paid form of non-personal presentation and promotion of ideas, goods or services by identified sponsors'. In other words, advertisement is paid use of any channel of communication – radio, television, press etc. to identify, explain or to urge the use or adoption of a product, service or idea. Advertising presents the most persuasive possible selling message to the right prospects for the product or service at the lowest possible cost. It is the means by which we make known what we have to sell or what we want to buy. Advertising is a powerful and difficult form of communication. It is a vital marketing tool which facilitates the selling of product, services and even the change of attitudes on social issues for the benefits of societies at large. When used for social purpose it is referred to as social advertising.

- The preparation of visual or oral messages and their dissemination through paid media for the purpose of making people aware of and favourably inclined towards a product, brand, service, institution, idea or point of view. - *Albert Fray*
- Advertising is the dissemination of information concerning an idea, product / service, to induce people to act in accordance with the interest of the advertiser. - *Frank Jetkin*
- The process of buying sponsor identified media space or time in order to promote a product or an idea. - *B.S.Rathor*
- Advertising is communicating with and influencing someone to do something– usually to buy a product or service and often to think something. - *Sushil Bahal*
- Advertising is the paid dissemination of information for the purpose of selling or helping to sell commodities and services or of gaining acceptance of ideas that may cause people to think or act in a desired manner. - *Subrata Banerjee*

Advertising presents a totally controllable message as the advertiser pays for the space, time and frequency of the message.

In a nutshell, advertising is a paid method of sales promotion. Its main aim is to bring a particular product or service, in non-personal way to the attention of buyers, to make them aware of the need for the products and services and to persuade them to buy by influencing their attitude.

Need of Advertising

The following points explain the need of advertising;

1. ***Providing information:*** Advertising is needed in order to provide information, knowledge and education to consumers. It is needed for giving information about new products or existing products to consumers.
2. ***Facing competition:*** Advertising is needed in order to face market competition from identical products and rival producers.
3. ***Creating and maintaining demand:*** Advertising is needed in order to create new demand and new markets for the goods. It is necessary for maintaining the existing internal demand and also for promoting exports.
4. ***Intensifying salesmanship:*** Advertising is needed in order to widen the scope and intensify salesmanship. It creates proper background for effective salesmanship. Advertising makes salesmanship more effective and purposeful.
5. ***Creating curiosity:*** Advertising creates attraction, interest and curiosity in the minds of consumers. It creates attraction and desire in the minds of people much before the actual introduction of the product in the market. Advertising creates proper background for the introduction of new product in the market.
6. ***Popularizing product / service:*** Advertising is needed for making the product or service popular in the market along with its manufacturer. It gives publicity and popularity to the product and also to the manufacturing company.

Objectives of Advertising

In a marketing programme, advertising helps us in attaining a number of objectives as under;

1. ***Increasing sales:*** Advertising can persuade potential consumers to visit dealers, retail outlets, company run show room and distributors. It can help in communication of special consumer offer and contents to consumers to promote purchase.
2. ***Introducing a new product:*** Advertising helps in moving the consumer from awareness to purchase of a product while introducing a new product.
3. ***Creating awareness or an interest in product:*** Advertising helps in building awareness on the nature of the products being

offered to the consumers. It educates them on the benefits to be gained from the use of these products and tells why these products are better than competing brands. It can also be used to communicate the price and price changes, about warranties and guarantees.

4. ***Sustaining an established product in the market:*** Advertising plays an important role in enabling a product to continue to sustain its market share.
5. ***Increasing market share:*** Effective advertising can either win away the market share from the competing brand or expand the total size of the market to the advantage of the brand being advertised.
6. ***Creating favourable impressions:*** Advertising helps in building an image for a company by highlighting its values and the role it may be playing in meeting consumer needs or for that matter serving the society in any way.
7. ***Countering wrong impression:*** Advertising helps in countering wrong impression that competitors could create for a brand.
8. ***Promoting a public cause:*** Advertising can also be used in promoting a public cause. Today, there is advertising to persuade the parents to get the infants vaccinated or to change mindset of society towards gender equality.

Functions of Advertisement

1. Basic functions of advertisement

Advertising fulfils the following basic functions in society;

a. ***Marketing function:*** It serves a marketing function by helping companies sell their products. Personal selling, sales promotions, and advertising blend together to help market the product.

b. ***Education function:*** People learn about new goods and services or improvements in existing ones through advertising.

c. ***Economic function:*** Advertising plays an economic role. The ability to advertise allows new competitors to enter the business arena, competition, in turn, encourages product improvement and can lead to lower prices. Moreover, advertising reaches a mass audience, thus greatly reducing the cost of personal selling and distribution.

d. ***Social function:*** Finally, advertising performs a definite social function. By vividly displaying the material and cultural opportunities available in a society, advertising helps increase productivity and raises the standard of living.

2. General functions of advertisement

The purpose and tasks are set by the marketing plans for advertising. The ready purpose of advertising is to sell goods, services, or merely an idea through effective communication. The following functions are served by advertising –

a. ***Promotion of new product*** – Advertising helps in making entry of the new product / service in the market. It informs, persuades, reminds and influences consumers in favour of the products which may be goods or services.

b. ***Support to personal selling*** – Advertising move the product nearer to the point of purchase. Advertising makes favourable conditions for the salesman as many things are known to the audience due to advertising. Thus the personal sale is facilitated by advertising.

c. ***Brand patronage*** – Advertising has great effect on brands and companies. The advertising programme aim at consumer awareness and positive attitude. The buyers are induced to purchase and repurchase the product. Advertising create and retain brand preference and brand loyalty. Brand ambassadors help in this process.

d. ***Immediate buying action*** – Advertising attempt to obtain immediate buying action by audience. The different advertisement techniques like mail order advertisement, point purchase advertisement, direct action retail advertisement, price deal offers, last chance offers and special advertisement persuade the audience to take prompt action.

e. ***Presold goods*** – Well advertised brands help for presold goods. Buyers are pulled by sure customers and goods are sold without active help of counter sales force.

f. ***Dealer support*** – Big firms advertise extensively and intensively to support dealers and distributors so that they can assure accelerated distribution. Advertising create new markets for products which are sound and can easily fill the customer needs and desires. Mass marketing brings reduction in the cost of production as well as cost of distribution.

Thus advertisements help in sale and promotion of goods. Many times the new products are of good quality but they are not known to the consumer. Unless the consumer knows the merits of the product he will not purchase and use them. The consumer forum is analyzing the products in the market and helping the consumers in purchase of quality products.

Types of Advertising

Advertising is accepted as an important part of any social system, and a large majority of the population feels that it creates better products and services for society. Advertising messages continuously surround us. Our value structures, attitudes, and basic needs affect how one perceives these messages. The most common types of advertising are standard advertising and public service advertising. In addition, contemporary social issues and governmental regulations have induced new types of advertising. Among these are social responsibility advertising, counter advertising, advocacy advertising and image advertising. Let's examine the different types of advertising in brief.

1. ***Consumer advertising:*** Most of the consumer goods producers engage in consumer product advertising. Markets of pharmaceuticals, cosmetics, scooters, cars, soaps, etc., are using this type of advertising.
2. ***Classified advertising:*** It provides valuable information in fact manner for example information about the employment market such as 'vacancy', about health, about marriages, about change of names, about availability of various services such as tuitions, matrimonial advertisements etc.
3. ***Public service advertising:*** These are devised by non-profit organizations to answer a public need. Time or space for this type of advertising is provided free as a service to the public by the print or broadcast media. This type of advertising focuses on social issues like family planning, national integration, population and campaigns against alcohol, drugs and smoking. This type is also known as social service advertising or development advertising.
4. ***Industrial advertising:*** In this type, machinery and large outlays are involved. The descriptions in such advertisements are precise and technical; the appeal is down to earth and rational. The presentations are mostly factual since the target audience is technical personnel of industrial companies.
5. ***Institutional advertising:*** The aim of this type of advertising is the building up of a good public image or an institute. This type focuses on a company's activities in research, development and quality control, on its sponsorship of educational, cultural and sports programmes and on its consumer service and social service programmes. For example, advertisement of 'Achievements of government or the Indian Railways'.

6. ***Product advertising:*** It is designed to promote the sale or reputation of a particular product or brand. This type of advertisement concerns the product itself or some of its features, such as service, price or the quality directly associated with it.
7. ***Other types of advertising:*** These includes legal tenders, legal notices, public notices, political advertising during election time etc.

A. On the basis of orientation of the advertisement

On the basis of orientation of the advertisement towards product, service or institution, the type of advertisement is decided. In this context the advertisement can be broadly classified in three categories;

1. ***Product oriented advertisement*** – It is one which is aimed at promoting the sales of the products by providing product information to the prospective buyers and users of the product.
2. ***Service oriented advertisement*** – It is one which is designed to promote the ideas and services to the public, offered by any particular agency or organization.
3. ***Institutional advertisement*** – It is one which is published to create good will or a favourable attitude or image towards the business or the institution.

B. On the basis of purpose of advertisement

1. Product or non-product

A product is a good or service for which customers / clients will exchange something of value. The purpose of product advertising is to promote or sell the goods or services. Non product advertising is also known as institutional advertisement; the aim is not to sell something, but just to let the public know that such as organization exist. It constitutes institutional / corporate advertising in which the manufacturer / institution / advertiser tries to polish or build up its image or influence public opinion about an issue.

2. Commercial or non-commercial

If the idea is to make profit, advertisement is called as commercial, otherwise it is non-commercial. Most of the advertisements we see are commercial. Non-commercial advertisement is undertaken by government bodies, charitable institutions, religious organizations or by political groups. These advertisements have objectives such as support of civic ventures, asking for donation and vote for candidate, etc.

C. On the basis of demand

Primary demand and selective demand advertising

Primary demand advertisements stimulate the need for a particular category and the aim is not to sell the brand. Once the need for a particular category is established, selective demand advertising comes in picture; this adverting concentrate, on the particular brand of a product and try to promote it.

D. Direct-action and indirect-action advertising

Sometimes advertisements are designed to get the audience to respond immediately. These types of advertisement are called direct-action advertisement. Typically, these advertise carry a toll free number and sale-promotion incentives. Indirect-action advertisement is done primarily to build awareness of the product.

The advertisements could also be classified on other bases as;

E. **On the basis of geographical area covered :** (i) National, (ii) Regional (iii) Local

F. **On the basis of message :** (i) Product, (ii) Institutional

G. **On the basis of appeal:** (i) Rational, (ii) Emotional

H. **On the basis of audience:** (i) Consumer, (ii) Industrial, (iii) Trade

I. **On the basis of sponsorship:** (i) Individual, (ii) Cooperative

J. **On the basis of results intended:** (i) Quick action, (ii) Delay action

Layout or forms of advertisement

Depending on the media the form of advertisement varies. In advertising, the term 'media' refers to communicational vehicle such as newspaper, magazine, radio, television, billboards, direct mail and internet. Advertisers use media to convey commercial messages to their target audience. Advertising in print media constitutes a major part of any advertising campaign. Television advertisements are volatile and more expensive at the same time. Print advertisements are relatively less expensive, have good reach and relatively longer life. They could be stored and used later. Advertising in print media can be designed keeping in mind the kind of audience. The newspaper / magazine are expected to reach by. So in more than one ways, print media have an upper hand in an advertising campaign and it is essential to know about all the technicalities in print media.

Before deciding the form of advertisement, the theme or appeal of the advertisement is decided. The theme is the central idea around which the whole advertisement is created. The various appeals used in the advertisement may be to sell the goods or provide the services. The selection of theme is usually made by means of consumer study or research. The common appeals with which advertising messages can be effectively communicated are beauty, pride, health, comfort, economy, fear, emulation, distinction, love and affection, status and prestige, ease and safety in operation, speed in performance etc. The theme points out the most important reason or driving force leading to the purchase of a certain product / service. It is the heart and soul of the advertisement. The examples of advertisement can be seen on the television and on other mass media.

The form deals with proper and attractive physical arrangement for the best presentation of the message or sales communication. Every product, corporation, or non-profit institution eventually takes on a personality by means of which people visualize it, remember it, and describe it to either people. Creative people must, therefore, understand the process of putting the different elements in an infinite number of ways. When they start putting their creative strategy in action, they must know which part of the advertising should come first, the headline, slogan, picture, brand name and so on.

A visual lay out of an advertisement has the following elements;

1. **Visual:** Though you don't absolutely require a visual, it will help draw attention to your audience. Research indicates that 70% of people look only at the visual in an advertisement, whereas only 30% will read the headlines.
2. **Slogans:** It is a phrase or a catchy sentence that describes either the benefit derived from the product or one of the product's most important attributes. It consists of a single phrase by which an advertiser conveys an important idea which will presumably lead readers to remember and think favourably of his company. It may play either a principal or a subsidiary role. The other name for slogan is theme line / tagline. It sums up the advertise. It is argument for product or service. It is equally important as the headlines as it is not only used for advertising but sales people and the company employees. It is the component of the advertisement which has maximum recall value. They are expressed in few words with brand name and name of the company. If the slogans are attractive and original, they are remembered by the audience for longer time. They act in their subconscious mind and act as hidden persuaders in purchasing the product. Now a day, there are brand ambassadors for various

products. For example the slogan of LIC – *'Jeevan ke sath bhee, Jeevan ke bad bhee.'*

3. **Headline:** A meaningful group of words or a sentence used to attract the attention of customers is called headline. It is the most important component in the advertisement. It must present main theme or appeal in minimum words. More often it works hand in hand with the illustration or other visual elements. A second function of most headlines is to lead readers into the text. Not only does the headline attract and select but it invites the readers to step in and stay a while. Instead of telling the story, it tells just enough to arouse interest. Let the name of the product be used in the headline for better recall value. It should be clear, simple, short, snappy and attractive to hold attention of the audience and generate interest. It must touch the people who read it. It must influence the readers emotionally, by making them laugh, angry, curious or think. It should reflect the content of the advertisement and guide the audience properly. It may be a promise or a reward and induce prompt action. It should be printed prominently.
4. **Body:** It is also called as text or copy. It describes the message of the advertisement. It is the written portion and main body of words or the textual matter of an advertisement. The copy, regardless of medium, is basically an amplication of the headline theme. It will point out the selling points of the product where it can be secured, what benefits the product will offer, what is its price and how it is to be used. It gives all essential information as well as guidance to audience. It provides proofs of benefits to convince the audience. It educates, reminds and promotes action of the buyers. It explains and develops the ideas presented in the headline. Copy is major category which constitutes the text of an advertisement. If you have compelling arguments, make them. If you have persuasive facts, state them. But don't overwhelm the advertisement with information. Two strong arguments will make more of an impression than a dozen weaker ones. Finally, be clear, precise and honest. Any hint of deception will instantly detract from entire message.

Style of copy / body

Straight line copy – It is continuation from the headline and is written in the words of an anonymous / unacknowledged source. It identifies the needs of the people and arouse desire in them.

Narrative copy – It is used to tell a story in first or third person. It persuades the same selling point in different context. Almost everyone likes to hear an interesting story.

The story is one of the oldest forms of literature and is a reliable approach to communicate an advertising point. Among the common forms of narrative copy are the short story, the picture and caption, and the comic strip.

Dialogue – It is presented in the manner as if readers listen to the conversation of the characters that are in the advertisement. This makes the advertise more believable as the characters portrayed in the print advertise speak from their own experience.

5. **Illustrations:** It provides cartoon, drawings, picture, diagrams, symbol, photographs etc. for attracting attention, creating interest and arousing desire. It is the most effective way of communication of ideas at a glance. It demonstrates the use of the products through proper visual presentation. An illustration in an advertisement is only one step behind from the real product or service advertised. It is, therefore, closer to reality than the written language. These are usually the most important visual elements in any print advertisement. To make an illustration selective, you must make sure it has some relevance to your product or to the copy. This should be clear to the readers when they first note the advertisement or when they read the copy. For example – use of photographs of farmers who adopted new cotton variety and compared with non-adopter farmers in advertisement.
6. **Logo:** It is symbolic representation of a company. It can be picture, or symbolic, or combination of both. Sometimes its significance is so much that the advertiser relies on the logo completely. People judge the reliability of the product based on logo. In other words logo adds to the credibility of the product.
7. **Colour:** It is an important ingredient of an advertisement. Colour has great attention attracting power. Judicious blending of colours may evoke emotional reaction. Each colour has individual meaning and significance in communication of ideas. Nature reflects its beauty in colours. Red, green, blue and black are most popular colours used in advertisement. Colour contrast is most attractive and pleasing to the eyes.

The form of the advertisement is decided by the arrangement of the advertising elements. Thus the layout provides a logical, clear, unified presentation of the advertisement message so that the message is remembered and acted upon by the audience.

Chapter - 11

AGRIPRENEURSHIP AND INDIAN AGRICULTURAL PROCESSING INDUSTRY

Agriculture is the biggest occupation undertaken by majority of the people in India. However, entrepreneurship has not been adequately developed in agricultural sector. The main reasons for this may be that most of the Indian farmers are either illiterate or less educated and till recently farming was done in traditional manner as way of life rather than an enterprise. Therefore the efforts are being made by government and non-government organizations to develop farmers into agripreneurs by inculcating in them the qualities of an entrepreneur.

According to Bahal (2008) agripreneurship is defined as generally sustainable, community-oriented, directly marketed agriculture. Sustainable agriculture denotes a holistic, systems-oriented approach to farming that focuses on the interrelationships of social, economic and environmental processes. Simply stated, agripreneurship means entrepreneurship in the broad field of agriculture.

Need of agripreneurship development

Agriculture is a composite family occupation comprising many areas and sub-areas. Within each area, enormous numbers of commodities are available. Based on the needs, agro-climatic conditions and available resources agripreneurs can adopt some commodities and flourish on it. There is also much scope to improve production and productivity in agriculture. It is, therefore, essential to train the jobless agriculture graduates and farmers in agri-business management and also provide finance to develop their own business in agriculture. Agripreneurship is required because;

1. Increasing demand for quality food and organic both in India as well as abroad.
2. Increasingly available modern agricultural technology being developed through research.
3. Competitive advantage for many primary production activities in agriculture such as tropical fruits and vegetables, livestock rearing, aquaculture, rainfed farming etc.

4. Supportive policies of government, NGOs and financial agencies.
5. Willingness of the private sector to enter into agri-business at all levels of operation.
6. Changing consumer demand and retail revolution.
7. Adequate scope for exporting agricultural commodities in the era of globalization.
8. Fast growing food processing industry.
9. Meeting the food security demand of malnourished section of population.

Agriculture to Agri-business : Key issues

Policy issues: From absolute control and management of agriculture by government, today it is being opened to Public Private Partnership (PPP).

Production technology issues: From inputs oriented technology development for increasing production to practice based value addition being promoted.

Quality and certification issues: Demands from consumers for better quality has forced the government to establish regulatory mechanisms for quality control and certification.

Logistics and supply chain issues: Modern retail formats require efficient and appropriate supply chain and management facilities. *Example:* timely delivery, temperature controlled supply chain for perishables from producer to consumer etc.

Human resource issues: Conversion of agriculture to agri-business or agripreneurship requires human resource development. Human resource in agriculture is being developed through agricultural education and training.

Government Programmes for Agribusiness

Both the central and state governments have devised programmes for entrepreneurs to set up agribusiness in India. These may be classified as programmes falling under small-scale industries, khadi and village industries, small and medium industries, large industries, based on the capital required for investment. Apart from these, to harness the export market, Special Economic Zones (SEZs), Agri-Export Zones (AEZs) are being set up. Government is encouraging private investments in infrastructure development like cold storage chains, improvements in road, rail, sea and air transport systems.

The Ministry of Small Scale Industries and Agro and Rural Industries, designs and implements the policies through its field organizations for promotion and growth of small and tiny enterprises, including the coir industries. The Ministry also coordinates with other ministries / departments on behalf of the Small Scale Industries Sector (Amarnath and Samvel, 2008).

Agri-clinic and Agri-business Centers

The scheme is open to graduates in agriculture, horticulture, veterinary science, animal husbandry, dairy, fish culture, agricultural engineering, food technology, sericulture, forestry and other allied activities. Under the scheme, the unemployed graduates are provided start up training at designated training centres. The successful candidates can later apply for start up loans for any specified venture. The project can be taken up by the trained graduates either individually or on jointly / group basis.

These centers started by trained agriculture graduates provide expert services and advice to farmers on cropping practices, technology dissemination, crop protection, market trends and prices of crops in markets and provide clinical services for soil, plant and animal health, which would enhance the productivity of crops and animals. Agri-business centres are those which provide inputs, supply farm equipments on hire and relevant services. Agriculture graduates may choose from a range of potential areas.

EXIM BANK

The Export-Import Bank of India (EXIM) set up for promoting India's foreign trade has the following functions, segmented into three operation groups.

1. Overseas Investment Finance, which handles a variety of financing programmes for Export Oriented Units (EOUs), importers and overseas investment by Indian companies.
2. Project Finance / Trade Finance handles the entire range of export credit services such as supplier's credit, pre-shipment credit, buyer's credit, finance for export of projects and consultancy services, guarantees, forfeiting, etc.
3. Export Services Group offers a variety of advisory and value added information services aimed at investment promotion.

EXIM Bank offers a diverse range of financing services for the Indian exporter, including a variety of Export Credit facilities and Finance for Export Oriented Companies. EXIM Bank offers term loans for setting up new projects, and for acquisition of assets for modernization, upgradation and expansion of existing units. The Bank also extends 100% refinance to commercial banks, for term loans sanctioned by the lending bank to an export oriented unit.

EXIM Bank offers term loans to export oriented units for development of new technology to satisfy domestic and international environment and standards, and to help them develop and / or commercialize new product / process applications.

Aligning EXIM procedures with WTO norms: Production and demand of various items in the country and their prices determine from time to time the EXIM policy. Care is taken to prevent shortfall of essential items in the country and dumping of goods by foreign countries.

1. ***Employment oriented measures:*** EXIM policy should initiate measures which would help employment generation.
2. ***Diversification of markets:*** Exploration of new markets is an essential component of EXIM policy. *For example:* Focus LAC (Latin American countries) was launched in November, 1997 and Focus Africa was launched in April, 2002.
3. ***Making exports more competitive:*** EXIM policy should facilitate improvement of technology in the export sector and the development of an efficient infrastructure.

Special Economic Zones (SEZs)

Special Economic Zone (SEZ) is a specific area notified by an appropriate authority for the purpose of conducting economic activities like manufacturing, processing etc., exclusively for export. Infrastructural facilities like roads, water supply, sanitation, electricity, communication, banking, insurance etc. are provided in these areas so that the entrepreneurs can perform without any hindrance.

An SEZ is a trade capacity development tool, with the goal to promote rapid economic growth by using tax and business incentives to attract foreign investment and technology. The Govt. of India in April-2000 announced the introduction of SEZs policy in the country, deemed to be foreign territory for the purpose of trade operations, duties and tariffs. The policy provides for setting up of SEZs in the public, private, joint sector or by State Governments. SEZs generate employment, income and earn foreign exchange and boost up economic development of the country.

1. SEZ units would be able to import capital goods and raw materials duty free.
2. SEZ units shall be deemed to be foreign territory for the purpose of trade operations and tariffs. Goods going to the SEZ area would be treated as deemed exports.
3. SEZ units would be able to obtain products from the Domestic Tariff Area (DTA) without paying terminal excise duty.

4. SEZ units would be required to export the whole of their production. The basic rationale of the SEZ units is to provide a totally free atmosphere conducive for exports.

Policy Initiatives for Food Processing Industries

In order to promote investment in the food processing sector, several policy initiatives have been taken. These include:

1. Food processing industry has been declared a priority area. So, it qualifies for a number of fiscal relief and incentives to encourage commercialization and value addition to agricultural products.
2. Full repatriation of profits and capital is allowed.
3. Almost the entire sector is delicensed, freeing it from bureaucratic hassles.
4. Automatic approvals for foreign investment up to 100%, except in few cases, and also technology transfer.
5. Zero-duty import of capital goods and raw materials for 100 % export-oriented units,
6. Government grants given for setting up common facilities in Agro Food Park.
7. Vision 2015 adopted by the Ministry of Food Processing Industries envisages:
   (a) Trebling the size of the processed food sector.
   (b) Increasing level of processing of perishables from 6% to 20%.
   (c) Value addition to increase from 20% to 35%.
   (d) Share in global food trade to increase from 1.5% to 3%.

Food Parks

Food Parks are being set up in various parts of the country to play a facilitating role in integrating the food chain, by providing common facilities like cold storage, quality control labs, pilot plants, uninterrupted power supply, adequate water, effluent treatment plants etc. Such Food Parks will help create backward linkages with farmers and processors and also provide forward linkages by linking processors and the markets, besides facilitating the link between the processors and the Research and Development institutions and financial institutions.

Mega Food Park

Mega Food Park is an industrial park meant for the food processing industry only. The government has planned to set up 30 mega food parks to help India double its share in global trade of processed food by 2015, out of which 10 are under implementation in the first phase. These are -1. Shirwal (Maharashtra) 2. Ranchi (Jharkhand) 3. Chikmagalur (Karnataka) 4. Dharmapuri (Tamil Nadu) 5. Jalandhar (Punjab) 6. Haridwar (Uttarakhand) 7. Jagdishpur- Raibareily (UP) 8. Jangipur (West Bengal) 9. Chittor (Andhra Pradesh) 10. Nalbari (Assam) (North East).

The Mega Food Parks are envisaged to be a well defined agri / horticultural processing zone continuing state of the art processing facilities with support infrastructure and well established supply chain. The proposed scheme aims to provide a mechanism to bring together farmers, processors and retailers, and link agricultural production to the market so as to ensure maximum value addition, minimize wastage and improve farmers' income.

The Mega Food Park is designed to link the farmers with the retail markets with minimizing the intermediaries, thus helping farmers earn higher returns from their produce. These food parks will be spread over 10 to 100 hectares. Minimum size of 10 hectares shall enable smaller food processing units to start operation within the SEZ. The government will provide a subsidy of Rs. 50 crore per park to private investors and will have cold warehousing, grading centres and research laboratories.

The Mega Food Parks will create an integrated value chain from the farm gate to the consumer and will envisage complete backward and forward linkages along with common processing facilities.

Facilities and incentives provided by government for entrepreneurial development

1. Development of industrial estate areas for transportation, bank, insurance, electricity, training, etc.
2. Quick approval of project reports.
3. Development of training facilities through Small Industrial Development Organization, District Industrial Centre, etc.
4. Establishment of science and technology entrepreneurs park.
5. Consultancy assistance and consultancy fee subsidy schemes.
6. Organization of seminars to discuss over the problems of enterprise and economic development.
7. Govt. subsidy to tribal and rural areas to encourage industries and exports.

Government incentives directly available to entrepreneurs

Investment subsidy: Government is providing investment subsidy to entrepreneurs to establish more and more industrial units. These include capital investment subsidy, transport subsidy, power generators subsidy, special subsidy to women entrepreneurs, etc.

Other subsidies: Other subsidies available are export / import subsidies, tax subsidy, excise duty exemption, capital subsidy scheme for technology upgradation.

Soft loan scheme for educated unemployed youth: People falling among the age group of 18 to 40 years can avail benefit of the scheme. Under the scheme, an amount of Rs 50,000/- for sole trading and Rs. 75,000/- for partnership business will be given, which would be 10% of cost of fixed assets of the business. The rate of interest is quite low on these loans.

Agri-horti Entrepreneurship in India

Since the demand for organic vegetables and cereals is increasing globally, India, being the second largest producer of vegetables and the fifth largest producer of soya in the world, could establish her entrepreneurial skills and capture the untapped areas in horticulture. Entrepreneurs in India can take advantage of the cooperative route, contract farming and the poultry industry to capitalize on various ongoing government incentive schemes, which are not known to many involved in this business.

Horticulture, which includes vegetables, fruits, flowers, plantation crops, etc., has immense potential for entrepreneurship development. There are many other areas of entrepreneurship development of which one should be aware of. The corporate sector of India has already entered in many of these areas.

Agriculture Industry in India

Some facts about Indian agriculture industry

- India with arable land of 184 million hectares.
- India produces annually 90 million tonnes of milk (highest in the world)
- India produces 150 million tonnes of fruits and vegetables (second largest)
- India has 485 million livestock population (largest)
- India produces 204 million tonnes food grain (third largest)
- India produces 6.3 million tonnes fish (3rd largest) ,
- India has 489 million Poultry and produces 45,200 million eggs.
- India had the *'Green Revolution'*, the *'White Revolution'*, and the *'Blue Revolution'*, and now the focus is on the *'Pink Revolution'*.

- The Green Revolution had led to self-sufficiency in food grains, the White Revolution saw India occupy the Number One Position in milk production in the world, and the Blue Revolution brought about increase in fish production.

One can understand the agriculture or agro industry as the offshoot of agriculture. Development in the agricultural industry is crucial to make agriculture more profitable and generate more employment opportunities both in the levels of production and sales. Â The comprehensive development realized in the agriculture industry will help enhance both the physical and social infrastructure in the subcontinent. Commercialization and diversification of agriculture will further add up to the income of the farmers and give way to food surpluses.

One can see the agro-industry mainly consisting of the post-harvest activities including processing and preserving of agricultural products for the purpose of midway or final consumption. Around the world, it is well understood especially in the segment of industrial expansion that the whole essence of agro-industries is connected to agriculture production as economies are seen developing. It needs to be stressed that food is not only that grows in the fields, it also refers to a variety of products that are processed from the harvest. Over and above, in developing countries like India, agriculture industry is a very important arm of the manufacturing industry to build on the industrial capabilities.

Agriculture Industry can be widely grouped under the following kinds. Village industries comprise those that are run with minimal industries with very little machinery and a maximum of manual labor manufacturing products like papad, pickle etc. Small scale industries refer to those that are managed with a medium level of investment and machinery like oil mills and rice mills for instance. Large scale industries involve huge investments and extreme levels of automation producing commodities like jute, sugar and cotton products.

From the time of its Independence, agriculture has been the most important deciding factor of India's economy backbone of India's economy. Agriculture segment in India alone was responsible for not less than 18% of India's GDP in addition to offering jobs to about 60% of India's entire population. Services related to the agriculture industry in India could make up a bulk of 62% to the country's GDP. However, irrespective of its immense contribution to the GDP, it only employed 28% of the total population of the nation. The volume of exports has been increasing between 2001 to 2009, making up a 200% increase during this time. However, despite the commendable growth in agriculture and agriculture products mentioned above, there has been a fall in the total volume of exports of agriculture products from India since 2004 to 2009 from 12.7% to 10.23%.

On the whole, India is the second largest producer of food on the globe. Indian agriculture industry enjoys a huge potential with respect to canned food, frozen food, processed food, food grains, dairy products, fish, meat and poultry, the growth of which will never cease over all times to come. Also, areas like sea fishing, milk and milk products, meat, aqua culture and poultry are some of the agro based sectors that have demonstrated a significant growth over the past few years. Strong links between the members of the food supply chains, existing policies and the prevailing business environment are sure to take advantage of the potential opportunity available in the global market today.

Experts in the arena reveal that the Indian Agriculture Industry is poised towards a great revolution that will modernize the whole food chain since the total volume of food production in the sub continent is most likely to double over the next ten years period.

An Overview of Indian Agriculture Products Industry

Across a number of sectors in agriculture, India enjoys the credit of being one of the largest producers including milk and dairy products. Indian agriculture is renowned for its enormous product diversity. A great deal of export prospects for Indian agricultural products ensues from the fact that the agrarian commodities produced in the country are complementary in nature as against the western countries. India is at present at the brink of making a revolution that will eventually modernize the food chain since the country is bound to see the doubling of the total agricultural production over the next ten years.

Recent studies show that the total volume of turnover in the Indian food industry amounts to over Rs.250000 crore (US $ 69.4 billion), of which, the contribution made by value-added food products is not less than Rs.80000 crore (US $ 22.2 billion). Over and above, the Indian government has also sanctioned a number of proposals featuring joint ventures, collaborations with foreign firms, industrial licenses and fully export oriented units requiring an investment of Rs.19100 crore (US $ 4.80 billion. Out of this, the value of foreign investment will be more than Rs. 9100 crore (US $ 18.2 Billion).

The agricultural food industry in India is highly significant in terms of accounting for about the country's 15% of GDP and generating an employment opportunity to about 65% of the population. The consumer food segment has a top priority in terms of featuring several joint-ventures and foreign collaborations in addition to inviting foreign investment. Some of the attractive benefits in the Indian food industry that have allured foreign investors include aqua culture, deep sea fishing and milk and dairy products besides meat and poultry segments.

In the processed food segment, India exports a good volume of hides and skins. Though the country exports semi-processed items to a large extent, it also exports fully processed bulk shipments of coffee and tea without branding or packing.

A few years back a number of firms were struggling to market their packaged food products. But, these days, it has become very easy to establish a firm footing in the Indian market owing to a younger population, enhanced incomes, evolving technologies and a fast developing middle class, estimated to be more than 50 million households. On an average, an Indian citizen spends about 53 % of his income on food. There is a huge domestic market for processed foods in the country which is growing rapidly in tandem with the economy, which is at present worth about $90 billion. Experts state that the processed food manufacturers in India need to demonstrate a persistent stand besides adapting their products to the preferences of Indian culture.

A number of large companies including ITC, HLL and Nestle stepped into the Indian market long ago and deeply penetrated into the domestic market. There are a few observations to be learnt from these success stories that will help in capturing the higher end of the domestic market besides grabbing a fair share of the export market. The model proposed by experts based on these observations include the strategies including large scale investment and adoption of the latest technologies, Intensive marketing efforts combined with tie-ups with foreign firms and creating a brand name.

Dairy products make up a very important arena in the Indian agricultural industry endowed with a huge potential. Though the nation has made enormous progress in the production and processing of milk and dairy products over the past twenty years, the fact needs to be acknowledged that not more than 15% of all the total milk produced in the country is processed. Today, a sizeable Indian population suffers from diabetic or cardiac ailments. Availability of fat free milk, fat free curd and sugar free food products is not adequate. For instance, the country needs to increase the production of soya milk.

There is a huge potential for the export of fish and shrimp from the nation however, this segment suffers for want of facilities with respect to cold storage and modern processing. In a year, the country produces more than six million tonnes of fish, while the frozen storage capacity is for only about one lakh tonnes.

Another potential area with regard to exports in India is herbal medicine. Worldwide, there is a growing awareness regarding the herbal products that they do not have any side effects. The country has a vast experience in the arena having tried and tested a vast range of herbal products over its long tradition. What is however lacking in this arena are rigorous quality control procedures, proper packaging methods and a unique brand name.

An encouraging fact remains that the Indian government and a number of modern retailers have started addressing these critical issues by bringing about new laws on packaging and labeling procedures as well as pooling out greater investments in the supply chain. Indian Agriculture Industry has a long history of more than 10,000 years and has developed to modern industry with its own peculiarities.

Food Processing Industry

Following are the essential conditions for processing industry in Agriculture and different Foods. Irrespective of the nature of raw materials or finished products, the food processing industry should have the following basic conditions.

1. Food processing industry should be set up and run in a clean environment.
2. Availability of raw materials and infrastructural facilities, including trained manpower must be ensured,
3. All operations of food processing should be performed under clean and hygienic conditions.
4. Strict quality control all through, from collection of raw materials to finished product, must be enforced.
5. Enhance shelf-life of products, during which there shall be no contamination or deterioration of the product.
6. The food produced should be hygienic, wholesome and tasteful.
7. Suit the taste of consumers in the country and abroad (for exported products).
8. The produce should have good market demand and generate employment and income.
9. The endeavour should be not only economically viable but socially desirable.

An Overview of Food Processing Industry in India

The food processing industry is one of the largest industries in India, it is ranked fifth in terms of production, consumption, export and expected growth. Food processing is a large sector that covers activities such as agriculture, horticulture, animal husbandry, fisheries, etc. It also includes other industries that use agriculture inputs for manufacturing of edible products.

India is the world's second largest producer of fruits and vegetables. The huge production base needs to be supported by strong food processing industry. The major items of the Indian fruits and vegetable processing industry are pulps particularly of tomatoes & mangoes, ready to serve juices, canned fruits, jam, pickles, squashes, etc. Frozen fruits, dehydrated and freeze dried vegetables, canned mushrooms etc. are also becoming popular among Indian consumers. The fruit and vegetable processing industry in India is highly decentralized with a large number of units are in the cottage/home scale and small scale sector. According to Ministry of Food Processing the installed capacity of fruits and vegetables processing industry has increased from 11.08 lakh tons in January 1993 to 21.18 lakh tons in January 2006.

Inspite of the facilities, the level of processing has been as low as 2.20% in case of fruits and vegetables, in comparison to 60% in U.K. and U.S.A., 70% in Brazil, 78% in The Philippines and 80% in Malaysia. The level of processing in other commodities has been to the tune of 14% in milk, 4% in fish and 1% in meat and poultry. Thus India's contribution to processing of farm produces in the year 2004-05 has been only about 1 % at the global level and 2% at the national level.

As per the Annual Report of the Ministry of Food Processing Industries (MoFPI), the turnover of the total food market for 1999-2000 was approximately Rs. 25000 billion out of which value added food products comprised Rs. 800 billion. But a caveat here is that as per the classification of MoFPI, food processing also includes segments like the aerated drinks, mineral water and alcoholic drinks, which are not part of this study.

The perishable nature of food makes the industry highly conducive to local operations. Small firms dominate the lower segment of the food market offering unbranded products that cater to regional tastes. Barring a few companies like Hindustan Lever Ltd, most Multinational Corporations (MNCs) have restricted themselves to premium western products like jam, sauces and chocolates. Very few have addressed the mass market where growth prospects are the highest but consequently risks are also greater. The MNCs have little control over the price and quality of agricultural produce as intermediaries dominate the supply chain serving as effective proxies for the distribution infrastructure. Further, basic food products are less price sensitive so MNCs cannot afford to pass on the inefficiencies in the system to the consumer.

The estimate about the actual size of the food market in India is very sketchy in the fact that every source, which estimates these figures, has their own different set of assumptions as also segmentation. Thus to get an idea of the market size in India, we could possibly look at the total final private expenditure on food. During the year 1996-97, consumption was pegged at over Rs. 3480 billion and was forecasted to grow at around 4% per annum in the future.

India's food processing sector covers food grains, milk products, fruits and vegetables, meat products, biscuits, chocolates and cocoa products and malted food. Of the total value added food products market of Rs. 800 billion. India exported processed fruits and vegetables worth Rs. 3393 million. There is however a declining trend in processed food products exports owing to various issues including quality.

Within processed foods exports, the segment that seems to have the greatest potential is meat and poultry products. India ranks first in world cattle population and boasts of 50% of the world's buffalo population and one sixth of its goat population. Currently buffalo meat is in surplus in India and has the potential to become a major foreign earner if modern slaughter and storage facilities are installed.

Milk and milk products which includes infant milk food, malted food, condensed milk and cheese, is another promising segment within the sector. When the world milk production registered a negative growth of 2% in 1999-2000, India performed much better with 4% growth. The total milk production was around 78 million tons in 1999-2000. The manufacturing of milk products is concentrated in the milk surplus state of Uttar Pradesh, Punjab, Haryana, Rajashtan, Gujarat, Maharashtra, Andhra Pradesh, Karnataka and Tamil Nadu. The total production of milk products stood at 31 million tons in 1999-2000. Production of milk powder including infant milk food rose to 0.23 million tons in 1999-2000, whereas that of malted food was around 65000 tons. Cheese and condensed milk production stood at 5000 and 11000 tons respectively. Plants are being set up for producing lactose, casein and improved cheese varieties.

Grains could be another major export earner for India in the coming years. India's food grains production is now at around 200 million tons. All major grains – paddy, wheat, maize, barley, jowar, bajra and ragi are produced in the country. India is the second largest rice producer in the world, with a 20% share. In fact, Indian Basmati rice has always been in great demand in the international market despite recent problems on the patenting front.

It is the Indian middle class that holds the key to success or failure of the processed food market in India. This segment estimated at 250 million people comprises mostly dual income nuclear families with higher disposable income that are more receptive to ready-to-eat food. It is this segment, which has given the impetus to the food processing industry in India.

The biggest bottleneck in expanding the food processing sector, in terms of both investment and exports, is lack of adequate infrastructure.

Without a strong and dependable cold chain, vital sector like food processing industry which is based mostly on perishable products cannot survive and grow. Even at the current level of production, 30-35 % farm produce, particularly fruits and vegetables, is being wasted every year, because of lack of adequate storage, transportation, cold chain facilities, and other infrastructural support.

An overview of Indian Dairy Industry

The strongest growth of dairying is to happen in Asia, notably in China and India as declared by the World Dairy Situation 2007 by the International Dairy Federation. The notable growth points are:

1. India is and will continue to be the largest producer of milk in the world and produces 82-92 million tonnes of milk annually, which is 15% of the world output.
2. There are 201 million cattle and 94 million buffaloes in India.
3. Dairy / Livestock activity accounts for 7% of National income and 26% of total agricultural activity in India.

There is a large scope for processing of milk by women from chilling to preparation of following milk products.

a. Fluid milk: boiled, chilled, sweetened, flavoured
b. Concentrated products like kheer and rabadi
c. Fermented products like curd, lassi, *etc.*
d. Coagulated products like paneer, chhaina
e. Frozen items like ice cream, kulfi
f. Fat rich products - butter and ghee
g. Sweet / Confectionaries - burfi, peda, gulab jamun, rasogulla, milk cake, *etc.*

The production units of above product could be established by entrepreneurs individually or may be established on cooperative basis. AMUL, WARNA, etc., are the conspicuous examples.

An Overview of Fish Processing Industry in India

The fish and fishery products processing sector is mostly export-oriented, with very little product making its way onto local markets. The development of fish and fishery product exports from India has gone hand-in-hand with the development of fish processing sector. Historically, fish was dried and cured following traditional practices and sold mostly locally. During the 1960s and 1970s, however, there was rapid transformation of fish processing, as fish was marketed more widely, initially within the country itself and then to export markets. Freezing and canning were first introduced in the early 1960s. Fish and fishery product export from India grew steadily from 1960 to 1990 at an average rate of 10% per annum, or from 16,542 MT to 133,653 MT. Historically, Indian exports of fish and fishery products have been directed at 3 major markets: the European Union (EU), Japan and the United States, which collectively account for approximately 85% by value. Of this, Japan alone accounted for over 45%.

Indian exports of fish and fishery products are dominated by frozen shrimp. From 1992-93 to 2003-04, the number of freezing plants registered with the Marine Products Export Development Authority (MPEDA) increased by 34%, from 258 to 352 numbers. Over the same period, however, total processing capacity in the sector expanded by 295% from 3,150 tonnes per day to 9,296 tonnes per day. Thus, average plant capacity more than doubled from 12.2 tonnes per day to 26.4 tonnes per day.

MPEDA is also promoting good practices to address the muddy-moldy smell that is having an adverse impact on exports of shrimp. Good practice includes regular cleaning of ponds and holding shrimp in fresh water for two or three days prior to harvesting. Work is also proceeding on a Good Aquaculture Practice (GAP) manual based on the responsible shrimp farming practices denned by FAO. (Source: MPEDA)

Status of Indian Farm Machinery industry

There has been a progressive shift from Draft Animal Power (DAP) to mechanical power in Indian agriculture because DAP and manual labour were not sufficient to cope with the workload of intensive agriculture. The use of mechanical power is becoming indispensable for making an optimal use of other resources and timely completion of various operations under intensive farming. Hence, mechanization has become a very crucial input to further the development of agriculture. Power operated equipments like tractor drawn mould board ploughs, cultivators, disc harrows, seed drill, threshers, plant protection equipments etc. have been accepted by the farmers in most of the states. Self-propelled and tractor operated combines, reaper harvester, potato and groundnut diggers are also commercially available.

Recently due to acute shortage of labour, the farmers have been shifting towards farm mechanization. Owing to small and fragmented lands when individual farmers are not able to purchase farm machinery, they are forming co-operative groups and are purchasing farm implements collectively.

The manufacture of farm machinery in India is quite complex which comprises village artisans, small scale industries, State Agro-Industries Corporations, and organized tractor, power tiller and engine manufacturers. The organized farm machinery manufacturers deal with - Agricultural tractors, Power tillers, Agricultural tools and implements, Combines, Reapers, Tractor parts and accessories of agricultural machinery, Sprinkler, Earth moving machinery and parts, Diesel engines, Rice processing machinery, and machines for Dairy and Food industries.

In addition, there are about one million village craft persons who handle bullock drawn implements and equipments, such as - Steel plough, Cultivator, Puddler, Sowing devices, Cane crusher, Bullock carts etc. They also manufacture and repair indigenous hand equipments such as spade, sickle, nirani, khurpi, sabal etc.

Diversification of agriculture needs introduction of new machines and the trend among the farmers to use increasingly larger tractors will vastly expand the scope of custom hiring of farm equipments because in future, multi-farm use will be the only way to keep the operating cost of farm equipments at a reasonably low level. The farm equipments which have done well in the field in recent years are listed here, and they have good potential to be used on large scale in future. Zero till seed-cum-fertiliser drill, Pulverising roller-cum-puddler, Rotavator, Sub-soiler, Hole digger, Ridger planter, Multi-crop seed drill, Sugarcane planter, Sugarcane stripper, Sunflower thresher, Tractor operated high capacity sprayers, Combine etc. (Following Chattopadhyay, 2007).

Equipments for drip irrigation, control of temperature in nurseries, refrigerated vans and cold storages for cold chain are getting popular. In fish trade, mechanized boats and trawler with refrigeration facility are already popular.

Changes in Food Processing Technology

All food processing involves a combination of procedures to achieve the intended changes to the raw materials. These are conveniently categorized as 'unit operations', each of which has a specific, identifiable and predictable effect on a food. Unit operations are grouped together to form a 'process'. The combination and sequence of operations determines the nature of the final product.

Industrialization, urbanization and consequent changes in lifestyle of the people have brought in changes in food processing technology. To have a futuristic outlook, some of them are mentioned here.

1. In contrast to earlier years, consumers no longer require a shelf-life of several months at room temperature for majority of their foods. Changes in family lifestyle and increased ownership of freezers and microwave ovens, are reflected in demands for foods that are convenient to prepare, are suitable for frozen or chilled storage, or have a moderate shelf-life at room temperature.

2. There is now an increasing demand by consumers for foods that have fewer synthetic additives, or have undergone fewer changes during processing. These foods more closely resemble the original raw materials and have a 'healthy' or 'natural' image. The manufacturers have responded by reducing or eliminating synthetic additives from products, particularly colorants and flavours, and substituting them with natural or 'nature-equivalent' alternatives.
3. The manufacturers have also introduced new ranges of low-fat, sugar-free or low-salt products in nearly all sub-sectors. New products are supplemented with vitamins, minerals and products containing organic ingredients.
4. Improvements to food quality have also been achieved through changes in legislation and administrative restrictions to control diseases like mad cow disease, bird flu, etc.
5. Increasing investment in capital intensive, automated process to reduce labour and energy costs.
6. There has been a change in philosophy from quality control, achieved by testing final products, to a more sophisticated approach to quality assurance, which involves all aspects of management.
7. Global sourcing of raw materials and ingredients has been a feature of some industries from their inception, for example, spices, coffee, cocoa etc. but this has expanded too many more sectors, to reduce costs and ensure continuity of supply.

The changes in technology have been influenced by a variety of factors: substantial increases in the costs of both energy and labour, by public pressure and legislation to reduce negative environmental effects of processing, particularly air or water pollution and energy consumption.

Food processing equipments now have increasingly sophisticated levels of control to reduce processing costs, enable rapid change over between shorter production runs, to improve product quality and to provide improved records for management decisions.

Microprocessors are now almost universally used to control food processing equipments. The automation of entire process, from reception of materials, through processing and packaging to warehousing are aimed at. This requires a higher capital investment by manufacturers but, together with improved quality assurance, reduces production costs and wastage. It increases production efficiency, uses less energy and often fewer operators, and generates increased revenue and market share from products that have higher quality.

Chapter – 12

COMMUNICATION SKILLS

The researchers have identified a big number of attributes of entrepreneurs, but the precise identification of entrepreneurial talent is difficult and remains elusive. Similarly the variety of the skills have been identified as essential skills for an entrepreneur. A few important of them are as follows:

1. Opportunity searching skills
2. Managerial skills
3. Fiscal skills
4. Marketing skills
5. Technical skills
6. Public relation skills
7. Problem solving skills and
8. Communication skills

All these skills are equally important. However, communication skill occupies central position for the reason that it provides a basis for other skills. The communication process is life and blood of the human individuals in general and of the entrepreneurs in particular. Entrepreneurs acquire and perform the other skills through communication skill.

Meaning and definition of communication

The word communication is derived from the Latin word *Communis* meaning common. This implies that when we communicate, we are trying to establish 'commonness' with someone through a message. Then communication is a conscious attempt to establish commonness over some idea, fact, feelings and the like, with others. In essence, it is a process of getting a source and a receiver tuned together for a particular message or a series of messages.

Following are a few definitions of communication;

- Communication is a process by which two or more people exchange ideas, facts, feelings or impressions in ways that each-gains a common understanding of meaning, intent and use of message. - *J. P. Leagans (1963)*

- Communication is anything that conveys meaning that carries a message from one person to another. - *Brooker*
- Communication is the process of sending and receiving messages through channels which establishes common meanings between a source and a receiver. – *Van den Ban and Hawkins*
- Communication is the process by which information, decision and direction pass through a social system and the ways in which knowledge, opinion and attitudes are formed or modified.

  - *Loomis and Beegle*
- Communication is imparting or interchange of thoughts, opinion or information by speech, writing or signs.

  - *American College Dictionary*
- Communication is the process by which messages are transferred from a source to receiver. - *Rogers and Shoemaker*

Process of Communication

According to Rogers and Shoemaker communication process consist of a source (S) sends a message (M) via certain channels (C) to the receiving individual (R), which causes some effects (E) i.e. changing the existing behaviour pattern of the receiver.

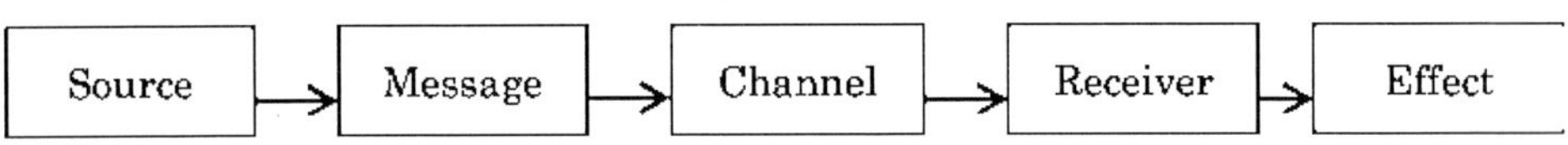

Figure 10: SMCRE model of communication.

This model is popularly known as SMCRE model. Brief description of each the elements is given below:

1. Source: It is point of origin of the message. The person who starts the process of communication is also called communicator. The communicator decides what message is to be sent, how to treat it, what channel to use and which receivers to reach.

2. Message: A message is the information that communicator whishes his audience to receive, understand, accept and act upon. The message may be information, instructions or orders.

3. Channel: It is anything used by a communicator of message to connect him with the intended receivers. In extension, these are the methods used by the extension worker. It may be a letter, meeting, radio, television or newspaper through which the communicator reaches the receivers.

*4. **Receiver:*** Receivers are the intended audience to messages. They are the consumers of messages. The communicator should identify and aim his message towards his intended audience.

*5. **Effect:*** Effect is the response by the audience to the message received by them. This may be some kind of mental or physical action. The effect should be viewed as a product of the communication process.

Importance of Communication

In management of an enterprise, the entrepreneur has to play the variety of roles in developing and managing an enterprise. He plays his role through communication. The functions of communication and the role of entrepreneur are similar and overlapping. The entrepreneur has to inform his colleagues and clients. He has to command and control the workers in enterprise. He has to influence and pursue the people and he has to integrate efforts of the people to bring harmony in the organization. All these functions are performed through communication. Communication is therefore a basic process for running an enterprise.

COMMUNICATION SKILLS

There are five types of the communication skills required to be mastered by the entrepreneur as below;

1. Writing skills
2. Reading skills
3. Speaking skills
4. Listening skills
5. Non verbal skills

*1. **Writing skills:*** The entrepreneur is required to make the correspondence to the suppliers, technicians, consumers and many such other people for running and managing his business. He has also to produce the technical and non-technical reports for the use by the public, financing and supporting organizations. The entrepreneurs should remember the 8 Cs of written communication;

a. ***Correctness*** – Accuracy in writing about any aspect of enterprise is a basic requirement of any good writing. The wrong reporting should be carefully avoided, so as to create faith among people about the enterprise or the company.

b. ***Clarity*** – The writing should be simple and clear to those groups of reader for whom it is being done.

c. ***Conclusions*** – Writing should be brief and precise. It should lead to clear conclusions.

d. ***Context*** – The situation in which and for which the writing is done, should always be kept in mind.

e. ***Content*** – Content should be interesting and useful to the readers.

f. ***Capability*** – This refers to reading ability, knowledge, habits and literacy level of readers. The writing has to be done in such a way that it matches to capability of intended readers.

g. ***Courtesy*** – There should be a personal touch whenever required in writing so that the readers get tuned to the entrepreneur or his representative.

h. ***Completeness*** – The piece of the writing or a report should be complete in itself. It should serve the purpose for which it is written.

The very basic aspect of written communication is that the purpose of writing should be clear to the writer. Secondly, he should choose the correct format depending on type of letter or report. Style of writing as per requirement should be followed. He should follow writing specifications of a kind of report, letter, advertisement, etc. Lastly, the entrepreneur cannot be justified for incorrect language and grammatical mistakes under the pretention that he is a technical person and a businessman.

2. Reading skills: The entrepreneur should have the reading skills as good as that of the writing skills, because he has to read the letter, reports and instructions from the supporting organizations. Effective reading is an art. Rapid reading skills should be developed, so as to save the time as entrepreneur is busy in several activities. The speed is affected by three factors. These are the number of fixations on each line, the duration of each fixation and general tempo of the effort. Skill of speedy reading has to be developed.

There are different reading styles like word reading, sentence reading and concept reading. A high degree of comprehension and experience is required for concept reading. The speed of reading also depends upon familiarity with the subject matter. Unfamiliar subject matter takes more time for reading. While skipping and skimming may do in case of familiar subject.

3. Speaking skills: The speaking skills for entrepreneur are necessary for conducting meetings, organizing group discussion and making presentations in seminars, workshops, etc. The entrepreneur is required to conduct meetings or participate in meetings as well as present his project before the audience. For oral communication to be effective, he should establish good rapport with audience. Show positive attitude, have pleasant appearance, use personal language, show enthusiasm and honesty.

Oral communication refers to not only the use of spoken language but also facial expression, gestures and voice manipulations in terms of intensity, pitch, tone etc. An effective oral communication should be able to achieve comprehension, understanding, perceived utility, validity and a lasting impression of the message in the mind of the listener. In oral communication skills, interpersonal relations assume great significance.

For presentation in a lecture form, entrepreneur should prepare his lecture carefully considering following points.

a. Spell out purpose in clear terms.

b. Collect adequate material.

c. Organize material in introduction, body and summery.

d. Add appropriate illustrations.

e. Read material thoroughly and prepare well.

f. Ensure that all teaching aids are ready.

While presenting, the entrepreneur should follow the following things:

a. Voice modulation – The pitch of the voice should neither be too low nor too high.

b. Generate interest.

c. Change in speed of delivery.

d. Combination of the auditory with visual presentation.

e. Support talk with gesture, posture, movement of hands.

f. Mix the eyes with audience.

g. Keep audience alive.

h. Follow good mannerism like saluting, thanking, etc.

i. Be in good dress and pleasant looking.

j. Pausing for emphasizing a statement or name.

k. Avoid following points in lecture method

- Don't walk up and down the available space.
- Don't show back to audience.
- Don't talk in same high pitch.
- Don't talk to window or roof.
- Be honest to admit the lack of knowledge about an idea.
- Avoid compulsive habits like playing with buttons, ring, pen, repeating some words very often.
- Don't stand between visuals and learners.

Good speaking skills increase the chances of success of an entrepreneur.

4. Listening skills: The entrepreneur should be a good listener in order to acquire knowledge, skills, right attitudes and other virtues. While taking a part in communication in a group one should not sit back in a relaxed manner. A good listener's mind is alert which reflects through his face and postures. He also shows interest by questions and comments. There are four different levels of listening skills as;

a. To make sense out of sound, i.e., to distinguish speakers words.

b. To understand what he is saying?

c. To segregate facts from fancy, i.e., to evaluate a statement.

d. To understand imaginatively other person's point of view.

Oral communication generally suffers due to poor listening on the part of the receiver. Studies show that only about a fourth of what has been heard is retained after two days. So, listening is a poor link in the chain of oral messages. Developing and improving the ability to listen is imperative in oral communication. It is important to train oneself to listen attentively. Listening well is a matter of correct attitude. Since listening efficiency is poor, considerable oral information is lost even in the process of transmission. One has to keep his eyes open. It is well said that open eye is the best indicator of open heart. However, it is a fact that listening for longer is a difficult task. Till for effective listening skills, one has to develop persistence in listening. Good listening is necessary for good speaking.

Listening skills include

a. Get prepared to listen to the speaker patiently. Recognize that one could gain something by listening. This could hold attention better.

b. Develop a positive attitude. This would enable one to look for and find something useful or valuable in the speaker's presentation.

c. Listen with an open mind. This involves respecting the other person's point of view. Do not deny or discredit a counter position. This would totally cut off listening.

d. Be focused, attend to the speaker with good rapport. This would help in eliminating listening problems to a great extent.

e. Concentrate on the messages being given by the speaker. Mentally look for and absorb the main ideas and thoughts of the speaker. Follow the speaker closely for grasping the full theme of the presentation.

f. Keep interruption/questions to a minimum. One needs to interrupt the speaker only if one has to. Otherwise this could detract from the focus on listening.

g. Use the non-verbal cues to clarify or support what has been heard. This would help in the consolidation of what has been heard.

h. Stop talking. To listen, one should stop talking. This is a useful guideline for better listening.

i. Listen with the 'mind' not only the 'ears.'

j. Basic knowledge in the subject will facilitate effective listening. If you are listening a lecture on spiritual empowerment for the first time on the concept of spirituality, one needs a minimum understanding of the subject for better listening.

Since the speaker and the listener are two partners in the process of interpersonal communication and sometimes they alternate their roles, listening plays an important role in interpersonal communication. If there is no effective listening, there is no communication.

5. Non-verbal skills: This includes messages expressed by non-linguistic means to attract and hold attention. One must make appropriate use of (i) voice, (ii) gestures, (iii) eye contact, (iv) smile, (v) enthusiasm, (vi) appropriate dress, (vii) time and (viii) distance.

It has been observed that non-verbal communication dominates the verbal communication because the people are eye minded rather than ear minded. The credibility of enterprise (company) and the entrepreneur counts very high in business and industry. Therefore, simply acquiring good communication skills by the entrepreneurs is not adequate. He should be honest in his dealings with people. Hence, the entrepreneur should know some fundamentals of communication.

Fundamental of Good Communication

Basics of communication revolve round as sense of humanity towards ones fellow being. HUMANITY in communication stands for;

H – Honesty, U – Understanding, M – Mutual respect,

A – Awareness, N – Negotiability, I – Identification

T – Trust, Y – Your creation

Use of communication process in an effective manner for managing an enterprise depends on knowledge and communication skills of entrepreneur. But this cannot be done in a day or two. The entrepreneurs have to consciously exert for acquiring communication skills for years together.

Chapter - 13

WRITING SKILLS

The entrepreneur is required to make the correspondence to the suppliers, technicians, consumers and many such other people for running and managing his business. He has also to produce the technical and non-technical reports for the use by the public and financing and supporting organizations. Therefore an entrepreneur should possess skill of writing. The entrepreneurs should remember the eight Cs of written communication namely;

(i) Correctness,	(ii) Clarity,
(iii) Conclusions,	(iv) Context,
(v) Content,	(vi) Capability,
(vii) Courtesy, and	(viii) Completeness.

The very basic aspect of written communication is that the purpose of writing should be clear to writer. Secondly, he should choose the correct format depending on type of letter or report. Style of writing as per requirement should be followed. Lastly, the entrepreneur cannot be excused for incorrect language and grammatical mistakes under the pretention that he is a technical person or a businessman.

Caution in writing a letter

Something that is in written form and once released cannot be taken back. Written communication is more concrete than verbal communication, with less room for error or mistakes. This presents written communicators with new challenges, including spelling, grammar, punctuation, even writing style and actual wording. Today's technology makes memo, letter and proposal writing much easier by providing reliable tools that check and even correct misspelled words and incorrect grammar usage. Unfortunately, these tools are not full proof and they require writers support, making his knowledge in this area important.

Basic tips for writing a letter

- Avoid the use of slang words.
- Avoid using abbreviations.

- Stay away with use of symbols.
- Clichés should be avoided.
- Brackets are used to play down words or phrases.
- Dashes are generally used for emphasis.
- Care should always be taken to spell the names of people and companies correctly.
- Number should be expressed in words when the number is less than 10 or is used to start a sentence. The number 10, or anything greater than 10, should be expressed as a figure.
- Double quotation marks should be placed around any directly quoted speech or single quotation for directly taken from a source and around titles of publications.
- Keep sentences short and simple.

Letter writing hints

- When writing letters, it is best to address the letter to an individual.
- When beginning the letter with a personal name, be sure to end it with an appropriate closing, such as 'Sincerely yours'.
- If you cannot obtain an individual's name, consider ending it with a more generic (less personal) closing, such as 'With kindest regards.'
- For normal business letters, your letter should start with an overall summary, showing in the first paragraph why the letter is relevant to the reader. It is not a good practice to make the reader go past the first paragraph to find out why the letter was sent to them.
- The body of the letter should explain the reason for the correspondence, including any relevant background and current information.
- Make sure the information flows logically, ensuring you are making your points effectively.
- The closing of the letter is the final impression you leave with the reader.
- End with an action point if any, such as 'I will let you know the exact date of next meeting.'
- Write a letter with personal touch and warmth.
- Follow the particular format and style for particular type of a letter.

Careful proofing of the letter

- Check letter thoroughly when it is completed.
- Use both the grammar and spell check on your computer.
- Have both a printed dictionary and thesaurus nearby to double check everything.
- Make sure the documents are clear and concise. Also ascertain that no part of written communication is amenable to be misinterpreted.
- The written matter should not raise unanswered questions or fail to make the point cut down on the number of words used, if needed.
- Written communication should be well organized. Each idea should proceed logically to the next.
- Make sure that written communication is easy to read, contain the necessary information, and using facts where needed.
- Avoid information that is not relevant.
- Again, outline the course of action you expect, such as a return call or visit.
- Close appropriately, making sure to include your contact information.

TYPES OF LETTERS

There are a variety of letters required to be written by entrepreneurs like business letter, enquiry letter, orders letter, etc. Following the Desai and Rai (2008) there are various types of letters.

[A] Business Letter

A business letter has a distinctive structure and appearance. Each of the parts has a specific style, position and function which has been established by custom and is followed because it is convenient for handling and processing correspondence in an organization's office.

Physical appearance of the letter

Though a letter is a written communication, its visual appearance cannot be ignored. A letter makes a visual impression before it is read, and the first impression is often decisive. Everyone is attracted and impressed by a good, presentable appearance. The appearance of a letter depends on layout of the letter, i.e., arrangement of different parts of the letter.

Paper of the letter: It must be of good quality. Most companies prefer to use pure white paper, with black print for the heading. However, coloured letter heads create good impression and bring out certain important elements. The envelope must be of the same quality paper with the same type of print as the letterhead.

Typing: Letter must be neatly typed without cancellations or corrections. The letter should be placed in the centre of the page, leaving appropriate margins on the sides and at the bottom. The margins are usually one-and-a-half inch on the left, one inch on the right, and at least one inch at the bottom. If the letter continues on next page, it must have the same margins as the first page. The printers and typewriters must be maintained well and cleaned regularly.

Paragraphing: It is not necessary only for pleasing physical appearance of a business letter but it adds to meaningfulness and understandability of the letter. It is necessary for breaking up a composition into readable, logical and progressive units. It is restful to the eyes to see a good amount of white space between the reading materials. In a business letter, paragraphing is important for appearance as well as for composition. Usually, a letter has three to four paragraphs: a short opening paragraph which may have only one or two sentences; one or two longer middle paragraphs which carry the main message of the letter; and a short closing paragraph which may be only one sentence.

Folding: The letter must be neatly folded otherwise it makes a poor impression. The size of folded letter must be such as to fit in the envelope leaving enough space for cutting the envelope open.

Address: Address on the envelope is typed halfway down and one-third in from the left in order to ensure that the post office franking or stamping does not efface it. It is typed in block form. Its appearance is as important as that of the letter as it is seen before the letter. Note that the practice of writing *To*, is outdated,

Format of a business letter

For usefulness and meeting requirements, certain parts of the letter have become essential. The main parts of the letter, apart from the text of the message are: Heading, Date, Inside address, Salutation, Complimentary close and Signature. In addition to these, the following details are included, according to the requirement of each letter: Reference number, Delivery mode, Subject line, Attention line, Enclosures, Courtesy copies, Sender's and Typist's initials. Each part has its own purpose.

Heading: It includes the company's; (a) name and business, (b) emblem, or monogram (c) postal, telegraphic, e-mail addresses, and (d) telephone, telex, and fax numbers.

Most companies prefer to have letterheads with a streamlined appearance. The printed portion does not occupy more than one-fifth of the sheet. It is printed in a simple attractive design in a single colour, usually black. The printed sheet is called letterhead.

Scientific Publishers (India)
5-A, New Pali Road, P.O. Box 91
Jodhpur – 342 001 (India)
Tel.: 0291 2433323, Fax.: 0291, 2613449
e-mail: info@scientificpub.com

Ref.... Date...

Date: Every official document and the business letter must be dated. The date provides reference in further correspondence and in following up the issues dealt with in the letter. The date is placed two line spaces below the heading, on the right hand side. In full block and NOMA styles, the date is at the left margin. The name of the month is spelled out in full; the year is written in all four figures. The figure of the year is separated from the date and month by a comma. Any one of the following styles may be used for writing date;

11 January, 2012

January 11, 2012

The practice of writing ordinal numbers as 1st, 2nd, 3rd, 4th, is not used now a days.

Reference Number: Reference number should be given on the left side at the level with the date. The words *Ref. no.* may be printed on the letterhead. Some companies have both, *Our ref. no.* and *Your ref. no.* printed on the letterhead, and give both numbers in their letters. The reference number gives the letter number, file number and other filing details. A letter is filed and can be located by its reference number.

Inside Address: It is the name and address to whom the letter is written. It is typed in block form; only in very few old-fashioned letters, the indented form is used. In the indented form the first line begins at the margin and the subsequent lines begin two spaces away from the beginning of the previous line; a comma follows each line. The indentions and the punctuation take more time to type, and also present a non-compact appearance. This style is now obsolete. In the block form, all lines begin at the margin and there is no punctuation at the end of any line. This style looks better and takes less time to type. The practice of writing *To* before the address may be avoided.

Block form	Indented form
Scientific Publishers 5-A, New Pali Road, P.O. Box 91 Jodhpur – 342 001 (India)	**Scientific Publishers,** 5-A, New Pali Road, P.O. Box 91, Jodhpur – 342 001 (India).

Attention Line: Attention line is used only if, letter is addressed to organization and you know the name of the person in the organization who is handling the issue or the subject. If a letter is directly addressed to a person in an organization, there is no question of attention line. The letter is sent to that person without delay because of attention line. The Attention line is placed between the inside address and the salutation, either at the margin or in the centre. It does not affect the salutation. The salutation must match the first line of the inside address.

For example:

> The Shree Furniture Manufacturers
> Mahatama Gandhinagar
> Station Road
> Aurangabad 431 005
>
> Attention: Miss A.C. Biswas
>
> Gentlemen,

Subject Line: Subject line gives a brief and quick indication of the subject of the letter. It directly opens up the topic to the receiver of letter and it helps him to decide whether pay attention to the letter immediately or not. It helps to classify and file the letter. The subject line is placed either above or below the salutation at the centre; in full block form at the left margin. When there is an Attention line, the Subject line is placed after the Salutation. It is introduced by the word *Subject:* or *Sub.*

For example:

> Dear Sir,
> Subject: Your advertisement no. 024/2011 in 'The Times of India

In more modern practice, the subject line is placed at the margin below the salutation, without any introductory word. It may be in capitals or initial capitals;

For example:

> The Manager
> The Shree Furniture Manufacturers
> Mahatama Gandhinagar
> Station Road, Aurangabad
>
> Dear Sir
>
> Providing quotations

Salutation: It begins at the left hand margin, two line spaces below the inside address. It is followed by a comma. In American practice it is followed by a colon. In full block style it has no punctuation mark and the salutation matches the first line of the inside address. If the letter is addressed to an organization, the salutation is plural: *Gentlemen or Dear Sirs or Sirs.* If the letter is addressed to a particular official mentioned only by designation, e.g., "The Sales Manager" the salutation is singular: Sir or Dear Sir. If the official is a lady, naturally the salutation is *Madam* or *Dear Madam.*

It is quite common, now, to address business letters to an official by name and designation, *e.g., Mr. M.P. Mehta, Sales Representative, or Ms P.P. Desai, Unit Coordinator.* In this case, the salutation is *Dear Mr. Mehta* or *Dear Ms Desai.* Ms can be used before a lady's name as it stands for both Miss and Mrs.

Complimentary Close: The complimentary close is written two line spaces below the last line of the text of the letter. It is placed on the right and is followed by a comma. In full block style and NOMA style, it is placed at the left margin and has no comma. The first word begins with a capital letter. The most common form is: *Yours faithfully;* Other common forms are : *Faithfully yours, Yours truly, Truly yours.* If the addressee's name is used in the salutation, the complimentary close is *Yours' sincerely,* or *Sincerely yours. Respectfully* is used only when writing to a high public official.

Signature: Signature gives authenticity to the letter. Also it gives a personal touch. The signature is placed just below the complimentary close. The name of the signatory is typed in brackets three or four line spaces below to leave sufficient space for the signature, and the designation / title is typed below the name. The complete signature includes the name of the company, either above or below the name and designation of the person who signs. The company's name is not given here in the signature if it is already in the heading. The following styles are in common use:

Yours faithfully, (Rupesh Patil) Marketing Officer XYZ Company	Yours faithfully, XYZ Company (Rupesh Patil) Marketing Officer

Per pro (Per procurationem) signature is used when the person who signs is not personally liable but has been given power of attorney to sign on behalf of the person or company who will be legally responsible. The words per pro or P.P. are put before the name of the person or company on whose behalf the letter is sent.

Reference Section: The reference section mentions enclosures, names of parties to receive copies, the type of delivery service to be used, and initials of the persons who dictated and typed the letter. Most of the details are for the convenience of the dispatch section. They are placed at the left margin after the signature.

(a) Enclosures: Enclosures are related documents sent with a letter. They may be bills, receipts, CDs, reports, cheques, quotations, brochures, price-lists, etc., which have been mentioned in the letter and are sent with letter. They may be folded and placed behind the letter in the envelope. If there are many enclosures, they are numbered and placed in the order of their serial number. The enclosures are indicated in the letter by writing *Enc (s) or Encls:* at the left bottom corner of the letter, after the signature. If there is more than one enclosure, the number is also indicated, e.g., *Encls: 2.* Sometimes the enclosure is named, *e.g., Encl:proforma invoice.*

(b) Courtesy Copies: When a matter concerns to more than one partly, a copy of letter related to the matter is sent for information to the other parties who are concerned. This is indicated by writing:

c.c.: Mr. M.G. Sharma

Or

c.c.: 1. Mr. M.G. Sharma
2. Sales Representative, XYZ Company

(c) Delivery Service: The type of delivery service to be used is indicated as Air Mail, Registered Mail, Express Delivery, Hand Delivery, Courier, Fax, etc. Dispatch section of the company sends the letter to concern parties accordingly.

(d) Initials: The dictator's and the typist's initials are placed at the bottom. This is for future reference in the sender's office; the person who

prepares or dictates a letter is not necessarily the person who signs it. The letter may be dictated by a senior clerk or a junior officer; it will be signed by the manager or a person who is authorized to sign for the company. The initials may be shown as: *PSW:rd* or *CRTgt* or *MRO/cs* or any such style.

Postscript: Postscript is a bit of writing containing not more than three lines, added to the letter after the signature and after enclosures. It is signed again, without the complimentary close, by the same person. It is used for conveying a small bit of information which is not a part of the main message of the letter. It is an extra or unrelated point and is often written in hand at the time of signing the letter. Postscript should never be used for adding something which was forgotten while preparing the letter. The idea of using PS to add a forgotten point is outdated with word processors. Even if a manual typewriter is being used, the letter must be typed again. The postscript can be used to add a friendly personal note to a formal letter; it may be just a line, but it can successfully establish personal contact by referring to some common experience or by making a friendly personal enquiry.

For example:

- *PS. It was nice meeting you at the launch last Tuesday.*
- *PS. How's your new venture in Mumbai coming up? Wish you the best of success.*

The postscript can be used effectively in a sales letter to impress an 'action incentive' sentence on the reader's mind.

- *PS: There's an early bird prize for the first seven orders.*

It can also be used to make an impact at the end of a 'stunt' collection letter.

- *PS: Do clear your account before the fast approaching festival season,*

Styles of Layout

Layout of a letter means the design in which the different parts of the letter are placed on the letterhead. The parts are placed in the same order from top to bottom in all forms or styles; the variation is in the indention, physical arrangement and the paragraph styles.

1. Indented Form: This is the oldest style and is now outdated. In this form, the inside address is typed in indented style as shown earlier, and every paragraph begins three to five spaces away from the left margin. The indention causes the letter to look uneven at the left margin. Besides, it takes more time to type because of the indenting. This style is not used today.

1 January, 2012

Indian Paper Mill Corporation,
22, Shivaji Nagar,
Mumbai.

Dear Sirs,

As per your price list, please supply the stationary as per the particulars attached with this letter within the 15 days. The payment of the bill will be within after delivery.

Yours faithfully,
Office Superintendent

2. Full Block Form: This is the modern style of layout of a business letter. There is no indention from the left margin at all; every line, including the date and the complimentary close begins at the left margin. The address has no punctuation at the end of the line. The salutation and the complimentary close do not have a comma at the end. There is double line space between the parts and between the paragraphs. Typists have an advantage in this form as it takes less time to type because it requires no change of margin and has less punctuation. Some companies do not like this form because it looks heavy on the left, and rather blank on the right. Quick reference to date and to signature is not possible in this form unless the file is fully opened.

1 January, 2012

Indian Paper Mill Corporation
22, Shivaji Nagar
Matunga, Mumbai

Dear Sirs

As per your price list, please supply the stationary as per the particulars attached with this letter within the 15 days. The payment of the bill will be within after delivery.

Yours faithfully

Office Superintendent

3. Modified Block Form: Modified block form eliminates the shortcomings of the full block style by keeping the date and the complimentary close on the right in their usual position. The inside address is in block form. The salutation and the complimentary close are followed by a comma. All the paragraphs begin at the left margin, and there is double space between the paragraphs. This is the most popular form, as it has many of the advantages of the full block form without its disadvantages. Its appearance is balanced and neat.

1 January, 2012

Indian Paper Mill Corporation
22, Shivaji Nagar
Matunga, Mumbai

Dear Sirs,

As per your price list, please supply the stationary as per the particulars attached with this letter within the 15 days. The payment of the bill will be within after delivery.

We shall be placing similar orders with you if we find that your products and services are of a high quality.

Yours faithfully,

Office Superintendent

4. Semi-indented Form: This form is also called Semi-block form. It has the inside address in block form but the beginning of every paragraph is indented. The date and the complimentary close are on the right side. The salutation and the complimentary close are followed by a comma as in the traditional style.

1 January, 2012

Indian Paper Mill Corporation
22, Shivaji Nagar
Matunga, Mumbai

Dear Sirs,

As per your price list, please supply the stationary as per the particulars attached with this letter within the 15 days. The payment of the bill will be within after delivery.

We shall be placing similar orders with you if we find that your products and services are of a high quality.

Yours faithfully,

Office Superintendent

5. Hanging Indention Form: In this style, the inside address is in block form. The date and complimentary close are placed on the right hand side. The beginning of every paragraph is at the left margin, but the subsequent lines of every paragraph are indented three to five spaces. The margins for the first line and subsequent lines can be set on a computer, but it takes more time to type on a manual typewriter. This style is adopted for sales letters, brochures and other documents in which striking appearance is useful. The opening line of every paragraph 'hangs out', and can be used for highlighting important points. If this style is used, the opening words of every paragraph should be important and worth highlighting.

1 January, 2012

Indian Paper Mill Corporation
22, Shivaji Nagar
Matunga, Mumbai

Dear Sirs,

Please supply the stationary as per particulars attached with this letter within the 15 days at rate quoted. The payment of the bill will be made after delivery.

Similar orders will be placed with you if we find that your products and services are of a high quality.

Yours faithfully,

Office Superintendent

6. NOMA Form: This is the most recent layout style. It is recommended by National Office Management Association of America and therefore termed as NOMA. It has been accepted in the UK by the Institute of Office Management. It has most of the features of Full Block form. However some special features of this form are: (i) it has no salutation and no complimentary close; (ii) the subject line is in capitals three lines below the inside address; (iii) numbered items of a list begin at the left margin, but if there is no number, the items are indented five spaces; there are no full-stops at the end of items; (iv) the writer's name and title are typed in capitals in one line below the space for signature; (v) the typist's initials are in the left bottom corner. This style is not popular as it is too far removed from the familiar, conventional styles. Some people feel that this style saves time and sensibly omits meaningless salutation and close; but most people feel that this omission makes it impersonal, distant, and lacking in feeling.

1 January, 2012

Indian Paper Mill Corporation

22, Shivaji Nagar

Matunga,

Mumbai 400 004

Sub: ORDER FOR SUPPLY OF STATIONARY

As per your price list, please supply the stationary as per the particulars attached with this letter within the 15 days. The payment of the bill will be within after delivery.

We shall be placing similar orders with you if we find that your products and services are of a high quality.

PRATIK PATIL

OFFICE SUPERIDENTENT

Psk

[B] ENQUIRY LETTER

Enquiry letters are sent to the firms for supply of quotations or price list for some goods or services. Sometimes enquiry letters are sent for acquiring certain information. The direct approach is used in enquiry letters to save time. The letter is short; formalities are not required, and you can get straight to the business.

For example: to ask for certain information, you can begin with any one of the following question;

- *Could you give us the following information?*
- *Please send us the following information:*
- *I would appreciate receiving the following information from you:*

Indicate the purpose for which the information is needed. In some fields, the information is very vast and only some of it is needed by different users; for example, a dealer in computers will be able to give useful information and a quotation only when he knows the buyer's specific need. The letter may begin with an explanation of the need as in the following examples;

- *We intend to set up a computer laboratory with capacity for 32 students to work at a time. The room in which the laboratory is to be set up measures 5 x 7 metres.*
- *We have been awarded a contract for the construction of a bridge across River Mula.. The work is to be completed in three years. We would appreciate receiving the following information from you:*

A series of direct questions may follow this opening paragraph. End the letter with a friendly comment which will leave a sense of personal relationship. Avoid old-fashioned endings like *'Thanking you in advance'* or *'an early reply would be appreciated.'* It is more effective to end a letter with an informal sentence like the following;

- *We could discuss this personally, if you could meet me in my office next week.*
- *If you could get this data to me by Monday, I shall be most grateful.*

The following content of the letter illustrates these points.

> Dear Sirs,
>
> During a recent Exhibition held here, I saw some samples of your new granite tiles for flooring. I would like to make use of them for the interiors I am designing for a new housing complex. Please give me the following information:
>
> (1) In what sizes, colours and designs can you supply the tiles?
>
> (2) Are the tiles likely to be affected by rising moisture?
>
> (3) Is any special processing of the under-flooring required?
>
> I shall be grateful if you could send me answers to these questions by Monday, the 21th.
>
> Yours faithfully,

Summarily above letter contains following points:

1. Courteous request for information, giving its purpose
2. Statement of what is wanted (or a list of questions)
3. Request that the information may be given by a certain date

[C] LETTER OF OFFER

Always give prompt attention to an enquiry made by a customer either by letter or by telephone. In the letter of offer, make every effort to work it up to an order. Your letter must do a sales job. You need persuasive skill and good judgment in giving information.

Focus on the facts showing the customer the benefits of buying the product / service from you. If you send a catalogue, information folder, brochure, price-list or samples send a letter with it; give the relevant page numbers or models which will be suitable to the customer. Ask for an appointment to meet the customer for a discussion. It is not enough to say, *'We hope that the samples will meet with your approval.'* You must show the advantages of the product.

The letters below illustrate these points.

Dear Sir,

Thank you for your request for our catalogue and price-list of plastic chairs and desks. They are enclosed.

You will find models shown on pages 3-4 particularly suitable for your purpose. We have supplied these chairs and desks to many schools, both in Mumbai and in small towns around Mumbai.

The schools have found them highly satisfactory as they require very little maintenance and are easy to clean. Their pleasant appearance adds to the attractiveness of the surrounding. We shall be very happy to discuss your requirements personally. Please call us at any of the numbers given above for an appointment for our representative to meet you.

Yours truly,

Letter of voluntary offers

An entrepreneur may make offers to potential customers without receiving any enquiry from them. Credit may be offered to customers of long standing in order to create goodwill strengthen relations further. It is also offered to individuals or firms of high repute and known credit standing to win them as customers. The following is an example of a letter offering credit.

Dear Sir,

You have been buying from us for over five years now, and we feel that you would like to enjoy our credit privilege.

With credit facilities you can keep good stocks and pay for the goods only after most of the goods have been sold, and your profits made. You can also buy in large quantities and take advantage of bulk purchase discounts. We hardly need to enumerate further, the various benefits that credit customers enjoy.

Please fill in and return the enclosed card and we would be glad to open a charge account for you.

Yours truly,

Letter for quotations

For buying a product or service, one calls quotations from several suppliers and then selects the one most suitable to him. If the order is likely to be large or might lead to further orders, indicate it in your request; suppliers offer special terms to large buyers. Mention all details of your requirements when asking for a quotation; if it is for service like painting, renovation, air conditioning, or a machine whose working you would like to see, ask for a representative to be sent.

Following points need to be covered in a letter asking for quotations;

1. Polite request for quotation.
2. Details of requirements with quality and quantity of goods.
3. Place and time of delivery of goods required, carriage paid or carriage forward.
4. Any special request about mode of delivery or packing.
5. Suggestion that the volume of business will be large if the quotation is favourable.

Sirs,

We intend to purchase large stocks of notebooks for our newly opened Department of Stationery in our Departmental Store at Salt Lake, and invite you to submit your most competitive quotation.

If your prices and terms are favourable, and your goods and services of a high standard, we shall consider giving a live year contract for the supply of notebooks. Please send your quotation, carriage paid, within a week.

Yours faithfully,

Letter for providing quotation

When market conditions are normal, the usual quotation is given. If the customer seems likely to place a large order, an exceptional quotation may be given. A tabulated quotation may be enclosed with a letter. Point out the advantages of the quotation and the benefit of buying the goods.

Quotation letter should cover following points;

1. Thanking for the enquiry
2. Statement of prices and terms
3. Attempt to win the customer

> Sir,
>
> We welcome the opportunity you have given us to introduce some of our products to you. We have sent you samples of our food colours and flavour concentrates, The enclosed price-list will show you that our prices are competitive. We give 7% discount on orders of Rs. 100,000 and more, and 2% cash discount for payment in cash within 7 days of the date of the bill. An order form is enclosed for your convenience. We are looking forward to the beginning of a pleasant business relationship with you.
>
> Yours faithfully,

Letter offering a substitute

If the goods demanded by a customer are not available, one can offer a substitute. The customer may be interested in another product which satisfies the need. The substitute you offer must satisfy the purpose and need of the customer. It must be the nearest in quality and price; if it differs, explain why. A substitute must be offered with confidence. Also win the customer's confidence for the substitute product. Sending a sample with the quotation or offering a 'trial pack' may help gaining confidence of the consumer.

Explain briefly why you are not able to supply the product which the customer has asked for. Simply offering a substitute without justification may create mistrust in the customer.

For example

> Dear Sir,
>
> We thank you for your enquiry of July 21, about the supply of Ashoka Xerox machines. The machines which you have asked for are out of stock and not likely to be available for at least next four weeks. We suggest the Himalaya Xerox Machines which have become very popular and have a good reputation for trouble-free service. There is a range of six models which allows you to choose a different kind for each desk according to the needs. These machines also have the additional feature of scanning and storage of job.
>
> The prices of machine range are a little higher than the prices of Ashoka but Himalaya machines have a guarantee of five years and can use several types of paper.
>
> We would be happy to demonstrate the Himalaya machine on any day and time suitable to you.
>
> Yours faithfully,

Letter withholding quotation

If many a time it happens that customer repeatedly asks for quotations for comparing with other quotations and does not place any orders. In such a case, you may avoid giving a quotation for which you have to write tactful letter mention that the market is uncertain and you would quote only for a definite contract.

> Dear Sirs,
>
> Thank you for your letter of 29 July, asking us to quote our rates for printing 5000 copies of your catalogue. As the price of paper *is* fluctuating and has to be negotiated for every order, we are unable to give a standard quotation. We shall be able to give a quotation only for a firm order. Our rates and terms for the printing job are the same as we had quoted for your brochures in April. We shall be happy to negotiate the price of paper on your behalf as soon as we hear from you.
>
> Yours faithfully,

Bargaining letter

It often happens that final price is usually negotiated in a face-to-face discussion. However some negotiations of price may be done by written correspondence. On receiving a quotation, a customer may try to persuade the seller to offer better terms. If the order is likely to be large, the buyer may point this out to the seller to persuade him to give better terms. The buyer may also indicate that other suppliers are willing to give more favourable terms.

Following points need to be covered in such a letter;

1. Expression of thanks for the quotation.
2. Making specific request for more favourable terms and showing the benefit to supplier.
3. Suggestion that large or frequent orders will be placed if the request is granted.

Dear Sir,

Thank you for your quotation for washing machines. We approve of your model Nos. 120 and 121 and would like to place an order immediately for at least 10 machines, five of each model.

However, we feel that your carriage forward terms are high. We usually get our supplies on carriage paid terms, and request you to reconsider this. We are planning to open launderettes in several places, and will be placing further orders if we find your products and terms satisfactory. You will be compensated by the large orders we shall be placing.

As we are about to equip our newly opened launderettes, we would like to have your reply within a week.

Yours faithfully,

Seller's reply letter

In response to request made by the customer, it may or may not be possible to reduce your quotation. However in either case, try to win the confidence of the customer. Agreeing to reduce your price should not give the impression that the originally offered terms were not fair. Also see that it should not create misunderstanding that sales are so poor that you are eager to get an order at any cost. If you agree to the request, try to put some condition like offering a special discount for a large order. Discount may also be offered for a long-term contract. A request or a favour should be granted with dignity. However, it should be remembered that a grudging tone will not be liked by the customer.

If you accept the request, the content of the letter will be as follows:

Dear Sir,

We are glad to learn from your letter of 19 February that you have undertaken a project to establish launderettes in several places in the city, and we are happy to be able to cooperate you in the project. Since this is a long term project and there will be continuous business for quite some time, we are prepared to supply our washing machines on carriage paid terms as you have requested.

Our machines are guaranteed for 5 years and are backed up by our excellent post sales service. We undertake to ensure your complete satisfaction with our products. Our representative will visit you within four days to inspect the location and check the facilities for installation of the machines.

Yours faithfully,

If you intend to refuse the request, remember all the skills of handling and write the letter in following manner.

Dear Sir,

We have given careful consideration to your request that we should give a 15% discount on our washing machines. Unfortunately, our overheads have risen sharply in the last 10 months, making it impossible for us to reduce prices to this extent without compromising on our standards of quality.

Since we cannot do this, we suggest that we will deliver goods carriage paid on all your orders for five or more machines.

We are sure you will understand that our pricing is done after taking into account the cost of maintaining high quality and post sales services. We request you to accept this offer and give us the opportunity of doing business with you.

Yours faithfully,

[D] ORDERS LETTER

Entrepreneurs are required to buy product and services for which they have to place orders. Orders are routine letters and their effectiveness depends on the accuracy of the details and specification of the product. Most of the contents of an order can be tabulated. Similarly the customers buy products and services from the entrepreneur or his agents. Replies to orders are treated as opportunities to build up goodwill and relationship with new customers, uncertain customers, regained customers as well as long standing loyal customers. Taking interest in replying to orders goes a long way in creating goodwill. Before placing an order, get all the details of what you want to buy. Group the details under headings as follows:

Product, its specifications and quantity: Some products come in a variety of sizes, colours, models and so on. Many products are sold in packages of different sizes. Liquids and small products have packing of various quantities; capsules come in strips often or bottles of. Some products are available in different types of packing like pouch, bottles, tins. This should be specified in the order.

Identify the exact product and item you want by giving specifications of (i) size (of the product or the packaging), (ii) colour, (iii) catalogue number, (iv) model number, (v) unit price. If you have selected the item after approving or testing of a sample, you may reserve the right to reject the goods if the delivered goods do not tally with the sample. Mention the correct quantity required. Different products are sold in different measures: liquids in litres; solids in kilograms; dozens or hundreds; large products - like cupboards or refrigerators, in units. All these details need be correctly given in an order letter. All these details need be given in an order letter.

Packing: Packing a consignment of goods suitably for transport is the responsibility of the supplier. Buyer's instructions are needed only if the goods are to be specially packed for any special purpose, such as for export.

Delivery: The place, time and mode of delivery have to be indicated at the time of placing the order.

a. Place: Give the full address of the place at which the goods are to be delivered. You may want the goods to be delivered at your retail shop or at your warehouse or at your showroom.

b. Time: If the order is urgent and you require the goods within a certain time, give the date by which the goods should be delivered or otherwise they will not be received. When the question of time is very important, and you have no use for goods delivered later, you may reserve the right to reject the goods if they are not delivered within the specified time.

c. Mode: The mode of delivery is required only if there is any special requirement. For construction work being carried out at a far-off location, you may want cement to be delivered in lots at specific intervals. Mode of transport need not be mentioned unless it is carriage forward.

Settlement of account: Mention the agreed prices and terms and refer to; (a) discount or concession given (b) mode of payment of the bill.

Order forms: Details to be given in an order are easy to tabulate, and therefore printed forms are used for placing orders. Sellers prepare the forms with columns required for their products. This ensures that no essential detail is left out; the form makes it easy for the customer to place an order. There is arrangement for the required carbon copies in the pad. Regular customers are given pads of the forms. A set of forms is sent with a quotation to encourage a potential customer to place an order.

Drafting orders: After the prices and terms are settled by a quotation and discussion, and samples are approved, the buyer may place a trial order to make sure that the seller gives efficient service that is, proper packing, delivery of goods in time, prompt receipts for payments etc. A retailer may also place a trial order to find out the demand for a new product. Only a small quantity is ordered for trial.

Any of the following situations may lead to a first order:

a. Seller's samples and / or quotation are approved.

b. A bargain has been concluded.

   i. request granted

   ii. buyer convinced to order on offered terms

c. Buyer is willing to try an offered substitute,

d. Buyer intends to try a product in a new area

Following points may be included in an order;

1. Reference to previous communication (for first order) and request to send goods.

2. Details of required goods, instructions for delivery and reference to terms of payment.

3. Good will message.

> Dear Sirs,
>
> We have tested your samples of PVC pipe and found that No. 52 is suitable for our requirements. We are glad to place an order for 500 PVC pipe. Please deliver at our shop at Pune. Our cheque for advance payment is enclosed.
>
> The material supplied should be of the same quality as the sample and we reserve the right to reject material of any other quality.
>
> Yours faithfully,

[E] STATUS INQUIRIES AND RECOMMENDATIONS

Letters related to customer status inquiries and credit must be written with the greatest care and tact. The company has to get information about the credit standing of customers who ask for credit.

Sometimes company may have to tell a customer that credit cannot be granted.

A buyer's credit standing depends on four Cs i.e. his Capital, Capacity, Circumstances and Character. All these aspects of credit are taken into account before granting credit to him. Capital is the cash as well as the deposited and stock, the building, the machinery and other property that the buyer owns. Capacity is the person's skill and ability to use resources properly and run the business/ profession profitably. It is also a fact which has to be remembered that capacity of the customer may be partly affected by the line of business and the location of the business. Circumstances are outside the person's control. Political conditions, government's financial and economic policies and industrial relations in the region or in the whole country may affect any business or enterprise. Currently, computerization of business as well as open market, free completion, globalization is causing changes in many organizations and production enterprises. Circumstances can change and may either improve or worsen the standing of a business enterprise. Character is what really makes a reputation of a person. It consists of honesty, ability to work hard, and self-respect.

Source of credit information

The credit information of a customer (desiring seek credit from the selling agency or company) may be gathered from sources like references given by the credit applicant, sources mentioned by first references, credit agencies, merchant's chambers, trade associations, market gossips about the customer. However, the most important and reliable source is a bank which knows several things about its clients. The scale of operation of customers business is assessed from amounts paid out and deposited; cheques issued and received by the customer. Bank can also provide information about conduct of the customer.

[F] CLAIMS AND ADJUSTMENTS LETTER

In any business, there are likely to occur some mistakes and accidents. One may not get the goods and services as per prescribed. If things go wrong, one has to bring the faults to the notice of the supplier and see that they are corrected and adjustments are made. The prime purpose of writing a claim letter is to get correction and adjustment. It is desirable to make a phone call immediately to the supplier to inform the deficiency, and then confirm it with a letter. If the supplier is a reputed company, it is usually enough to point out the mistake in the execution of the order and leave it to the company to make the adjustment. When there is a mistake in the execution of an order the customer suffers inconvenience. The customer must ask the supplier to take steps to reduce the inconvenience and make adjustments. If the customer states clearly what mistake has been made and what inconvenience has been caused, the seller can take steps to reduce it. The customer can help the seller by suggesting what can be done so that things can be set right and adjusted as quickly as possible.

Courtesy is of great importance; resist the temptation to accuse the seller of carelessness, negligence, inefficiency and indifference to customers. Courteous mention of error will earn the gratitude of a self-respecting seller. Firmness, emphasis and insistence on your rights can be done without discourtesy. Courtesy requires that offensive and unpleasant words should not be used. Words like *dishonest, false, careless, unfair, useless, inefficient* are offensive and discourteous. It is better to use passive voice to talk of errors, and choose words tactfully. It sounds more polite to say, in passive voice, '*Our product was not packed with the necessary care,*' than to say '*You packed our product carelessly.*'

Reasons for writing complaints

A customer may have one or more of the following reasons:

1. Damaged goods
2. Delivered goods are not the same as ordered
3. Delivered goods are not according to the specifications
4. Quantity of goods is different from what was ordered
5. Goods are delivered at the wrong place
6. Delay in delivery of goods
7. A product does not function properly
8. Wrong billing or reminders for payment after you have paid the bill.

Your claim and your expectation of the seller's response will depend on whether you are a commercial consumer. As a commercial consumer you make adjustments, so that the inconvenience and the hindrance to business are minimized. There is more inconvenience than personal loss. You also have greater power to get adjustments and compensation because, as a regular buyer, you have a hold on the seller. As a consumer, you may suffer personal loss on account of the seller's fault or negligence. You will expect correction and compensation, but a small individual buyer has less power in the market to get grievances redressed.

Letter of complaint

Following points may be included in letter;

1. Reference to the order, its date, number
2. Clear description of the mistake

3. Clear statement of the loss caused, and the action the supplier should take to lessen the inconvenience
4. Request for adjustment and investigation

For example

> Dear Mr. Pranav,
>
> We had placed an order with you on 18th January for 50 mobile cover to be delivered by the 25th of this month. Unfortunately, we have not yet received them. As we have not received any information from you about the delay, perhaps something has gone wrong somewhere. Will you please let us know when we may expect delivery?
>
> Yours sincerely,

REFERENCES

Adhikary, M.M., Acharya, S.K. and Basu, D. 2008. Extension Strategies for Managing Agricultural Enterprise and Entrepreneurship. Agrotech Publishing Academy, Udaipur.

Akhouri, M.M.P., Mishra, S.P. and Sengupta, Rita 1999. Trainers' Manual on Developing Entrepreneurial Motivation, NIESBUD, New Delhi.

Amarnath, J.S. and Samvel A.P.V. 2008. Agri-Business Management. Satish Serial Publishing House, New Delhi.

Bahal, Ram 2008. Agripreneurship Development in India, in Dynamics of Entrepreneurship Development in Agriculture: Basics to Advances (eds) Dipak De and B. Jirli. Ganga Kaveri Publishing Hourse, Varanasi.

Charndrgi, D.M. 2002. Communication Skills for Entrepreneurship. Paper presented at the Winter School on Advanced Strategies for Successful Entrepreneurial Development for Women, 5-25 August, ICAR, New Delhi and University of Agricultural Sciences, Dharwad.

Chhabra, Vinesh 1998. Marketing of Products and Services. Paper presented at the National Training Course on Entrepreneurship Development in Agriculture, 8-19 December. Directorate of Extension, Ministry of Agriculture, New Delhi and Centre for Agricultural Technology, IARI, New Delhi.

Chole, R.R. 2010. Process of Entrepreneurship Development. Paper presented at the Winter School on Entrepreneurship Development and Management, 15 Sept-5 Oct., ICAR, New Delhi and Dr. Panjabrao Deshmukh Krishi Vidyapeeth, Akola.

Chole, R.R. 2010. Communication and Enterprise Management. Paper presented at the Winter School on Entrepreneurship Development and Management, 15 Sept-5 Oct., ICAR, New Delhi and Dr. Panjabrao Deshmukh Krishi Vidyapeeth, Akola.

Cole, A.H. 1949. Entrepreneurship and Entrepreneurial History, in Change and the Entrepreneur. Prepared by the Research Centre in Entrepreneurial History. Harvard University Press, Cambridge.

Desai, Vasant and Urmila Rai 2008. Entrepreneurship Development and Business Communication. Himalaya Publishing House, Mumbai.

Desai, Vasant 2000. Small Scale Industries and Entrepreneurship. Himalaya Publishing House, Mumbai.

Desai, Vasant 2000. Dynamics of entrepreneurial development and management. Himalaya Publishing House, Mumbai.

Drucker, Peter 1985. Innovation and Entrepreneurship, Heinemann London, Harper and Row.

Drucker, Peter 1970. Entrepreneurship in Business Enterprise, *Journal of Business Policy,* Vol.-I.

Grover, Indu 2008. Handbook on empowerment and entrepreneurship, Agrotech Publishing Academy, Udaipur.

Grower, Indu, Sethi, Nishi and Deepak Grover 2004. Handbook of Communication and Media, Agrotech Publishing Academy, Udaipur.

Gupta, Vinod 1998. Steps in Setting up a Small Enterprise. Paper presented at the National Training Course on Entrepreneurship Development in Agriculture, 8-19 December. Directorate of Extension, Ministry of Agriculture, New Delhi and Centre for Agril. Technology, IARI, New Delhi.

Holt, David 2006. Entrepreneurship – New Venture Creation, Prentice Hall of India Pvt. Ltd., New Delhi.

Jhamtani Anita, Sharma, J.P. and Rashmi Singh 2003. Trainers Manual on Entrepreneurship Development in Agriculture. Division of Agricultural Extension, IARI, New Delhi.

Leagans, J.P. 1963. The Communication Process in Rural Development. New York, Cornell International Agriculture Development Bulletin, Cornell University, Ithaca.

Maity, T.K. 2007. Value Addition in Vegetable based Enterprises. Paper presented at the Winter School on Advanced Extension Strategy for Entrepreneurship Development and Management, 4-24 December, ICAR, New Delhi and Bidhan Chandra Krishi Viswavidyalaya, Kalyani, West Bengal.

Mancuso, Josesh 1974. The Entrepreneurs Handbook. Vol I & II, Artech House Inc. USA.

Mondal Sagar and G.L. Ray 2009. Text book of Entrepreneurship and Rural Development, Kalyani Publishers, Ludhiana.

Murthy, C.S.V. 2002. Small Scale Industries and Entrepreneurial Development. Himalaya Publishing House, Mumbai.

Pant, J.M. 1998. Financial Analysis of Projects. Paper presented at the National Training Course on Entrepreneurship Development in Agriculture, 8-19 December. Directorate of Extension, Ministry of Agriculture, New Delhi and Centre for Agricultural Technology, IARI, New Delhi.

Rogers, E.M. 2003. Diffusion of Innovations. The Free Press, New York.

Rogers E.M. and Shoemarker F.F. 1971. Communication of Innovations. The Free Press, New York.

Sharma, J.P. 1998. Product Selection. Paper presented at the National Training Course on Entrepreneurship Development in Agriculture, 8-19 December. Directorate of Extension, Ministry of Agriculture, New Delhi and Centre for Agril. Technology, IARI, New Delhi.

Shiv Khera 1998. You can Win. McMillan India Ltd., New Delhi.

Singh, A.K., Lakhan Singh and R. Roy Burman 2006. Dimensions of Agricultural Extension, Aman Publishing House, Meerut.

Sridharan, S. 2002. Entrepreneurship Concept, Nature and Need, Paper presented at the Winter School on Advanced Strategies for Successful Entrepreneurial Development for Women, 5-25 August, ICAR, New Delhi and University of Agricultural Sciences, Dharwad.

Sur A. 2007. Dairy Industry a Backbencher in India's Globalization Journey. Paper presented at the Winter School on Advanced Extension Strategy for Entrepreneurship Development and Management, 4-24 December, ICAR, New Delhi and Bidhan Chandra Krishi Viswavidyalaya, Kalyani, West Bengal.

Viramgami, H.S. 2008. Entrepreneurship Development, APH Publishing Corporation, New Delhi.